Zero Point Energy Device (ZPED) for Health, Healing, Ascension and Mankind's Future

Zero Point Energy Device (ZPED) for Health, Healing, Ascension and Mankind's Future

Brought to You by Neological Technology
www.neologicaltech.com

By
Kosol Ouch, James Rink,
John Nelson, Koeun Noun Ouch,
Ailene Liquete Pelingon,
Elaine Lagrimas Liquete

E-BookTime, LLC
Montgomery, Alabama

Zero Point Energy Device (ZPED) for Health, Healing, Ascension and Mankind's Future

Copyright © 2011 by Kosol Ouch, James Rink, John Nelson, Koeun Noun Ouch, Ailene Liquete Pelingon, Elaine Lagrimas Liquete

All rights reserved. No part of this book may be reproduced or transmitted in any form or by any means, electronic or mechanical, including photocopying, recording, or by any information storage and retrieval system, without permission in writing from the copyright owner.

Library of Congress Control Number: 2011907476

ISBN: 978-1-60862-292-4

First Edition
Published May 2011
E-BookTime, LLC
6598 Pumpkin Road
Montgomery, AL 36108
www.e-booktime.com

Contents

What is a Neo? 7
Real Technology 19
Weekly Meditation 30
Com Uplink 33
About Us 34
Protocol Systems 35
Career Opportunity 58
Local Provider 64
Order 65
Neo Zenmaster IDL-64 68
Neo Zenmaster IDL-12 72
Neo Zenmaster IDL-4 74
Neo Link Bracelet 76
The Inventor 84
Remote Healing 86
Vocabulary Section 88
The User Experience 95
Kosol History 169
Conclusion 184
Pictures 187
Appendix 1 Neo Cube Flyer 196
Appendix 2 Brochure 201
Appendix 3 User Manual 205

What is a Neo?

Every person has the ability to heal themselves, the problem is emotional trauma, toxins in the environment, and genetic damage can force our internal self healing mechanisms out of equilibrium. To bring the body back into balance and positive health requires an integrated approach including healthy diet, detoxing of the body, and correcting blocked pathways of life force energy.

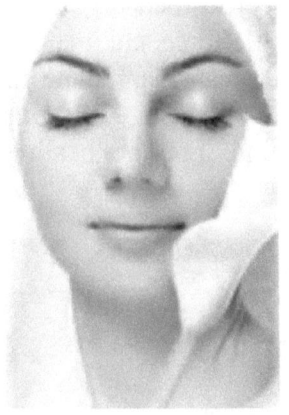

Neological Technologies recommends everyone eat a healthy diet and to detox their body as a positive step in the right direction. But our mission is to integrate the astral body. Just like with Yoga, Reiki, Acupuncture, Acupressure, Tai Chi, and crystals, a Neo can help you open up blockages of chi energy within the aura. But unlike these methods our Neo mind meditation machines can do so in a much faster, noninvasive, and easier way bringing integration to your mind, body, and spirit.

Neo Mind Meditation machines contain a built in torsion field generator which transmutes spinning fields of negative chi energy into positive chi energy, helping you release stress so that your body can begin to restore itself. Remember stress is the major reason why your body is unable to heal itself. When you reduce stress your immune system is enhanced and your pineal gland can open up activating your 3 to 12 strand DNA. There isn't any product on the market which can do this.

What makes a Neo so powerful is that this chi energy can help you tune into the consciousness energy field all around us. According to quantum field entanglement all parts of the whole are in all places at all times. Therefore if you can tune into this same frequency found in both sacred geometry, solfeggio frequencies, and Fibonacci numbers you can simply tap into a version of yourself that was healed in a different place or time.

This vibrational frequency is the very essence of consciousness technology, it is much more powerful then orgone or radionics. Call it what you like; prana, life force, chi, manna, aether; whatever you call it this device can manipulate it, bringing you positive health and well being to your subconscious and super-conscious mind. Let the healing integration begin.

SO WHY IS MEDITATION SO IMPORTANT?

Meditation has been used for centuries as a tool to bring the body back into balance and to tap into higher dimensions connecting you with God. Being in tune with this vibrational frequency is important because it helps unlock any blockages in your DNA preventing your body from healing yourself. Remember it's not the Neo doing the healing work, it is yourself. The key is to relax. It's that simple.

WHY A MEDITATION DEVICE?

Yogis, Ascended Masters, and Savants all have something in common they are all naturally in tune with the consciousness energy field. You can be as well, but it may take you 30 years of daily training to achieve that status. Who has time for that in this stressful day and age? Thankfully the Neo produces powerful torsion field energy accelerating this process from years to months. In fact after 256 hours of usage your life will be totally changed.

This technology is more powerful than anything on this planet because it is based on consciousness energy. Remember consciousness energy is found all throughout nature in and within sacred geometry. When you tune into this harmonious vibrational frequency you can tap into the universal divine collective consciousness, which is the very breath of god. This is why prayer, meditation, and being relaxed is so important in our lives because we can only tune into these higher dimensions when our body is in the right state of mind. Our bodies are scalar wave energy receivers which means we all have the ability to tap into the quantum energy field allowing you to manifest any of your desires. We are powerful beings, believe it!

"Therefore I tell you, whatever you ask in prayer, believe that you have received it, and it will be yours." Mark 11:24

HOW IT WORKS?

A Neo is a meditation assistance device. In order to feel the effect you need to meditate with it for 15 to 60 minutes per day. If you do not meditate with the device on a regular basis you will not gain the benefits as described. Their are two types of Neo's the Neo Zenmaster IDL-4 and Neo Zenmaster IDL-12 both units bring in chi energy into the body for integration and relaxation.

The 2"x2"x2" Neo Zenmaster IDL-4 requires 10 minutes to get to full power. If you meditate for less than 15 minutes a day you will need a more powerful device such as the Neo Zenmaster IDL-12. First time users will notice an increased sense of peace and relaxation. When the sensations end peace and love will stay with you during your conscious waking life. The main aim of the device is to destroy stress, but everyone will have their own unique variation of this experience.

The Neo Zenmaster IDL-12 is a powerful device in a small package measuring only 4"x4"x4". This unit powers on instantly and it's like your own personal monastery. This unit is similar to the Zenmaster IDL-4 except it's much more powerful and can be used for reanimating very sick individuals such as those in hospice care or nursing care. It works by drawing chi energy into your body so that it can relax and being the self healing process. Results will vary for each user, but over all expect to be more relaxed and calm.

The Neo Zenmaster IDL-64 is an extremly powerful device measuring 10"x10"x10". This unit is meant to be used by professionals to not only help reanimate the very sick, but it can also be used by reiki masters and energy healers, simply activate the unit with your left hand and use your right hand to send powerful amounts of chi energy. Yoga instructors can use it to help their clients in guided neo meditations to journey into other dimensions, and hypnotherapists can use the device to supercharge their therapeutic practice while helping their clients relax and go into trance further. This is also the same unit used on the weekly guided neo meditation show. The uses of this technology are unlimited.

THE EFFECTS

Some of the results you may notice after using a Neo include:

Increased Energy
Increased intelligence
Integrated Mind, Body, and Spirit
Enhanced Athletic Performance
Reversal of Addictions
Wealth Manifestation
Increased Longevity
Removal of negative entities from the aura
Enhanced intuition
Enhanced healing ability a must for reiki and energy healers
Balanced Chakras
Ascension and DNA repair
Spirit, Animal, and Plant Communication
Astral Projection
Lucid Dreaming
Access the Akashic Records
And much more!

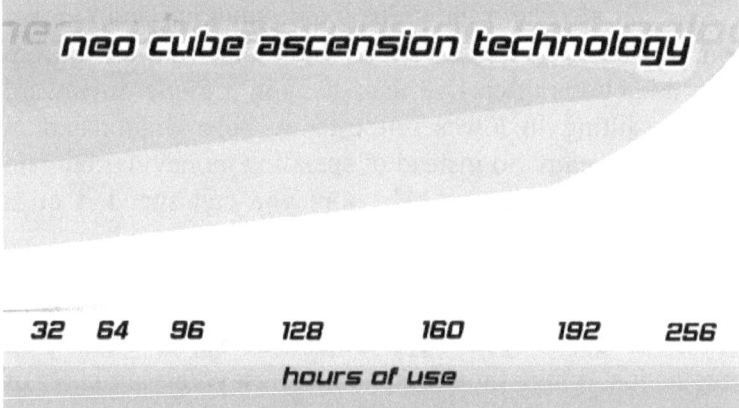

Charge up your body with Neo Cube Meditation Ascension Technology

New users will initially notice their intelligence increasing; synchronicities will increase in their lives such as attracting individuals who can assist in the ascension healing process. Some users may even manifest financial doors opening up. You should start to feel the changes taking place after 1 to 4 hours of use. It generally takes 32 to 64 hours of daily use to fully integrate yourself. At 256 hours of use, users will begin to unlock their dormant latent abilities.

Advanced users of a Neo have been able to reverse schizophrenia, bipolar, PTSD, and dissociation disorders. These disorders are so tragic as there is no cure and often require heavy doses of medication. A Neo can help these individuals integrate their minds quickly bringing them back into healthy emotional balance. One of our biggest clients are law enforcement officers and military personnel, these individuals have been exposed to so much trauma they often find it impossible to relax. If that describes yourself then be prepared your life is about to change.

A Neo can also help individuals quit smoking and recover from substance abuse addictions. Individuals suffering from these addictions often live their lives in a tragic downward spiral resulting in lower self esteem, poor health, and in some cases death. So instead of spending money to maintain these negative addictive behaviors you can spend it on a Neo and kick these habits once and for all.

The device can also be used to manifest the side effects of certain drugs. If you have cancer or suffer from pain imagine just telling the device to manifest your pain killer of choice to save money on medication. But always be sure to consult with your doctor before making any decisions when it comes to medication. Remember this technology works in conjunction with modern day therapies not in lieu of it.

The Neo is not just limited to mind integration it can also heal DNA damage, some users have even been able to reverse "incurable diseases" such as diabetes, autism, and even cancer.

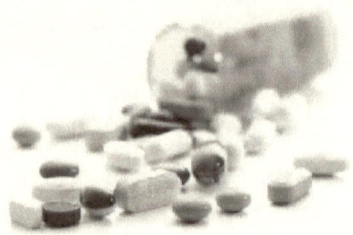

This device is great for elderly people because it increases chi energy into your body; chi energy is life force energy. As we age DNA replication errors reduce the level of chi energy brought into your body. Eventually this will lead to death. But a Neo can reverse this downward spiral slowing down aging and perhaps even reversing it if you continue using the technology for the rest of your life.

A Neo is also great for athletes as the increased chi energy will give you more power and strength. You can also instruct the chi energy to grow your muscles bigger muscles giving you an edge over your opponent. There are even some professional baseball players using this technology.

A Neo can also increase the healing abilities of reiki masters and energy healers. Simply activate the device placing one hand near the device and use your other hand to send healing energy. Not only is the device great for energy healers it can also be used as an income source for yoga practitioners and holistic healers. This technology can be offered as alternative therapy to help your clients who are just not getting better with the standard amount of herbal remedies, hypnosis, acupuncture, etc.

This technology can even remove negative demonic attachments from the body. Remember chi energy is the vibrational frequency of God, therefore anything not in tune with this frequency will not be able to stay around for very long.

INSTRUCTIONS

If you have a specific healing request then simply tell the device what you want it to work on. Don't worry about doing any chants, mantras, or visualizations, the device can tap into the consciousness energy field and will know exactly what it needs to do as long as you remember to operate it on automatic mode.

If you do not feel anything the first time, be patient, as it may take 4 hours of use before you notice any effect. This technology responds best to individuals with 3 strand DNA and up. Most people only have two strand DNA especially if you are new to meditation and have numerous health problems. In your particular case please use the device for a minimum of 32 hours before deciding if this technology works for you.

OPERATION AND USE

1. Only one person should use a machine at a time.
2. Turn off cell phones
3. Device must be operated in a dark room. Turn off all lights and close shutters, or place a cloth over your eyes to block out the light.
4. Be sure to drink some water if you are feeling dehydrated.
5. Take off your shoes and put both feet on the floor.
6. Now sit down in a comfortable chair. Do not lay down while using the device.
7. Place unit on a table or desk or a pillow on your lap.
8. Place both hands 6" inches from the device with your palms facing towards the device. Do not touch the device while in operating. If you accidentally touch it then repeat the activation protocol.
9. OPTIONAL: You may want to play some relaxing music such as a hemi sync or even a self hypnosis track while using the device.
10. Before using the device you may want to ask the Neo to help you on a specific issue. If you are unsure what to ask then go on to the next step.
11. Now close your eyes and recite the activation protocol. It can be done both vocally and internally.

BASIC ACTIVATION PROTOCOL

This is a general all purpose method, when you don't know what protocol to use start with this one. You should begin to notice a difference in 1 to 4 sessions. We recommend that you memorize this activation sequence before moving on to more advanced protocols.

Be sure to say the activation code properly as it will not work if you miss this step or do it incorrectly.

Recite "DEVICE ACTIVATE AND INCREASE"

Wait 30 seconds to three minutes in silence.

When you begin to feel tingling or pulsing sensations in your hands tell the device "DEVICE I TRUST YOU COMPLETELY TO HEAL AND INTEGRATE MY MIND, BODY, AND SPIRIT."

Wait for the stargate to appear (it should look like ring of light or a tunnel of light if you don't see it the first time be patient it could take a few minutes, you may also feel floating sensations) when you see it recite the following phrase "STARGATE I TRUST YOU COMPLETELY….. DEVICE ACTIVATE STAR GATE MODE AND INCREASE"

"AUTOMATIC MODE" – **always remember to say this!**

Now sit back and relax for 30 to 60 minutes. There is no need of a mantra or chant just sit back and let the device work on automatic mode for you. If you have specific request be creative for example ask it to inject nanites, serums, or enlarge your pineal gland.

Every 10 minutes or when you feel the device is slowing down recite "DEVICE INCREASE"

When you are done recite "DEVICE END SESSION"

With each session journal your experiences and visualizations in a notebook. Make a note of sensations, feelings, places you traveled.

This technology is easily the most worthwhile investment in health and clarity that you can ever make! Worth substantially more than a two week vacation. This is the law of attraction on steroids and then some. Now is the time to change your life. Now is the time to get a Neo!

And for a limited time save money while you relax!

LEGAL NOTICE - DISCLAIMER
Warning: Statements expressed within this book have not been evaluated by the Food and Drug Administration. Any and all information and/or statements found within this book are for educational purposes only and are NOT intended to diagnose, treat, cure, prevent disease or replace the advice of a licensed healthcare practitioner. Neological Technologies does not dispense medical advice, prescribe, or diagnose illness. Any views and ideas expressed are not intended to be a substitute for conventional medical advice or service. You agree that no responsibility or liability will be incurred to any person or entity with respect to any loss, damage, or injury caused or alleged to be caused directly or indirectly by the information contained within this book. If you have a severe medical condition, please see a licensed healthcare practitioner.

Real Technology

A NEOLOGICAL SOLUTION FOR OUR DIVINE BIOMIND

INTRODUCTION

The Neo is an interactive mind meditation machine and healing tool which allows any person to connect with their higher self, enabling you to consciously create whatever you want or desire in your life by simply navigating and interacting with the zero point energy field.

We all have the capability of becoming powerful ascended masters and avatars but due to trauma on the emotional, mental, physical, and spiritual levels we experience blocks expressed as illness and set backs in our evolutionary path. Changing this belief system can be accomplished through years of meditation and yoga, or it can be done technologically with the right interface.

According to quantum entanglement all pieces of the whole are in all places at the same time. Which means we are in fact a reflection of the universe in the macro form and we are the micro reflection of the universe in microform. In this holographic universe, each tiny part of the whole contains a fragment of all the other parts creating an infinite loop of unified oneness. It is this oneness also known as the collective consciousness that is the basis of all reality. Using meditation or with the right technology one can tap into this quantum energy field allowing you to manifest anything you desire.

Our brain is a biological quantum field device, when linked to consciousness energy it can become the most powerful computer on the planet. Everything in this universe operates on a vibrational frequency. When our mind is out of sync of this frequency, illness and disease result, but when you are in tuned you find there are no limits whatsoever. To tap into this frequency, you need to use a quantum field interface such as the Neo.

The average person only uses 5% to 10% of their brain. When we sleep we use a little bit more because our mind is in two locations at the same time. Individuals that use more of their brain are categorized as geniuses, autistics, and spiritual avatars. They all have something in common; they are able to link to the conscious, subconscious, super conscious, and collective consciousness mind. Because the average person cannot do this they often seek outward validations through religious cults and sects.

The Neo will allow you to bypass all these gatekeepers and open a direct dialogue between god and yourself by simply opening doorways of your mind in the quantum energy field. Tap the unlimited potential within your soul. Anything you want to know or experience you can. Teleportation, time travel, and clairvoyance are only the beginning of what we all are truly capable of. With 10 units you can make your own Montauk styled consciousness chair, but with 100 units you can travel the universe.

THE DEVICE INTERIOR

The Neo is designed to create and harness vibrational frequencies based on the principles of sacred geometry. This is the same technology used in the artifact found by Dr. Jonathan Reed.

Alien artifact is the same technology but enhanced with nanotechnology.

Our device does not use nanotechnology but the effects are still the same.

It consists of three parts:

1. Copper Cube
2. Copper Cones
3. Copper C.P.U. or Consciousness Processing Unit

The cube shape is significant as it represents the shape of everything in the universe. It is the perfect form of god. Every angle and shape throughout the universe in its base form looks like a cube; ranging from microscopic cells to large a black hole and even chi, electromagnetic, and electrogravitic energy. Only when a cube spins fast enough does it take on the form of a sphere. Remember the language of the consciousness energy field is geometry which is why a cube is able to tap into the vibrational frequency of the universe. A cube also has the same geometry of a tetrahedron, that is, it can shift into 144 variations which correspond to 144 frequencies and dimensions, which is the vibrational frequency of god.

Tetrahedron

The cones are embedded into the inner wall panels allowing them to collect and channel spinning fields of torsion energy or chi energy into the CPU unit located in the center of the device. To a clairvoyant this would appear as a spinning tetrahedron, or a 4 sided pyramid.

The core of the unit contains the CPU or Consciousness Processing Unit. This spherical shaped object contains multiple layers of copper hemispheres arranged according to the Fibonacci sequence and help create vibrational frequencies just like a musical instrument creates harmonic frequencies. This frequency resonates with the Schuman resonance, also known as the earth's heart beat and is amplified and radiated to any objects which are close by. The CPU unit and the Schuman resonance are both based on a Fibonacci sequences which corresponds to chakra's 1 through 7 on your body. As chi energy is directed from the tip of the cone into the CPU it passes through each layer of metal, creating a life force capacitor. Between each layer of metal is human hair which contains many trace elements; including gold, platinum, iridium, silver, aluminum, as well as copper.

In the inner earth cities of Agarthia and Telos the inhabitants actually use this same type of technology to meditate, but their units contain 144 group layers with each group containing layers of gold, platinum, silver, copper, and aluminum separated with layers of molecular plant fibers embed with human DNA, compare this to the Dr. Jonathan Reed artifact which has 18 group layers. These cultures have replicator machines and can produce the raw materials cheaply and on a nanoscale. Since we don't have that luxury, our units are made with a single copper layer but thanks to the trace elements within the hair they too carry some of these same frequencies of these rare and exotic metals. Each one of these elements carries its own vibrational torsion field frequency manifesting healing, paraelectric fields, as well as electromagnetic properties. The neo also produces all three types of these frequencies.

Healing - This is what makes you feel relaxed, calm floating, healing.

Paraelectric - This makes you feel warm all over, allows electricity to be made, just like monks who meditate in the cold. You feel tingling but warm.

Life Force - This means you are exposed to electromagnetic radiation that is converted into life force. This means you won't get tired, sick, and you will feel strengthened.

Our units contain both female and male human hair; because they are unisex they carry the personalities of both donors. Female DNA is good because you get the motherly instincts that kick in to prevent someone from using this technology for evil. Male DNA is good because they are hyper focused on getting the task done. DNA also serves as a A.I., artificial intelligence, interface which allows your brain to interact with the akashic field this is because DNA contains micro amounts of crystals which operate as scalar wave antennas allowing you to travel in between dimensions.

As chi energy is compressed into the center of the unit it eventually becomes super nova and opens up a star gate allowing you to draw in dark matter and dark energy into this third dimensional reality. 27% of the universe is dark matter, 70% is dark energy, and 4% is regular matter like you or I. The reason you can't see it; is because it is light that is so bright it appears dark to our eyes. It is the zero point, it is the 5th dimension, and it will lead you to ascension. This dark matter and dark energy then outflows into your hand chakras; syncing your body with the vibrational frequency of the consciousness energy field; helping to recharge and energize you; while making you healthier, intuitive, and more intelligent.

Each Neo contains symbols on the outside to guide the user on its function and use. (1) This is our company logo. (2) The circle on top of the pyramid shows you which direction the chi energy is flowing. Since the chi is spinning all around the unit there is no right or wrong way to hold the device. (3) The triangle symbol is a representation of the tetrahedron and (4) the circle inside the triangle means this unit is a pure chi outflow unit. A pure chi inflow unit takes energy from your body and may be harmful to your health which is why we don't carry this type of unit. (5) The letters "IDL" simply stands for interdimensional light, which is another name for consciousness energy. (6) The number located next to the IDL is the amount of copper layers within the CPU. So an IDL 4 would simply mean this unit has 4 layers of copper within its CPU which can be used to open up a star gate. The more layers within a unit, the more room there is to compress chi energy resulting in a more powerful star gate. (7) Is the model type. All Zenmasters are pure chi outflow units.

METHODOLOGY

To use the device sit in a chair in a dark room with both feet on the floor. Place the device on a table or pillow on your lap. Do not touch the device at any point during the session this will shut the unit off. Hold both hands on each side, palm side up towards the device. At this point you may want to tell the device what you would like to manifest in your life. It can be done either audibly or mentally as there are no wires. This is because the device interfaces using the quantum field effect, or the aura system surrounding the body. Then close your eyes and say.

"DEVICE ACTIVATE AND INCREASE."

After a minute or so you should begin to feel tingling or pulsing sensations in your hands, when you do, tell the device "DEVICE I TRUST YOU COMPLETELY TO HEAL AND INTEGRATE MY MIND, BODY, AND SPIRIT."

Wait for the star gate to appear. It should look like ring of light or a tunnel of light within your third eye. If you don't see it the first time be patient it could take a few minutes, you may also feel floating sensations. Now recite the following phrase "STARGATE I TRUST YOU COMPLETELY.....DEVICE ACTIVATE STAR GATE MODE AND INCREASE.......... AUTOMATIC MODE. "

The reason why we say star gate mode is because the pineal gland is a star gate to other dimensions.

At this point the Neo will take over and run on autopilot. If you do not say automatic mode the device will not know what to do and you will not get any benefits. Now simply relax for 30 minutes to an hour. You don't have to think of anything but if you do have a specific request feel free to ask.

For example you can ask for insight on certain issues in your life. Or you could ask to be injected with a healing serum that could either; integrate your mind, balance your emotions, heal your spirit, remove any pain, or repair and strengthen you.

Occasionally when you feel the effects wearing off and want to raise its power you may want to say...

"DEVICE INCREASE"

At the end of the session simply say...

"DEVICE END SESSION"

The more you use this device the smarter it will get because it records everything you say, so be careful what you try to manifest.

THE FINAL RESULTS

Because the brain is a vibrational transformer it can convert vibration frequency into sensation. You may start to notice different emotions expressed as visual, auditory, or tactile experiences. You may also notice yourself feeling really good. Once you begin eating chi energy you will become addicted because it's such a pure source of energy. In fact you will want so much chi energy you may reject all other forms of energy such as eating and sleeping. This just may become the new drug of the century.

You may also notice your latent psychic abilities will become enhanced. But one should understand there is no such thing as psychic energy as it's a natural ability that exists in all individuals and life forms. Anyone can harness this gift by simply tapping into the consciousness energy field through meditation.

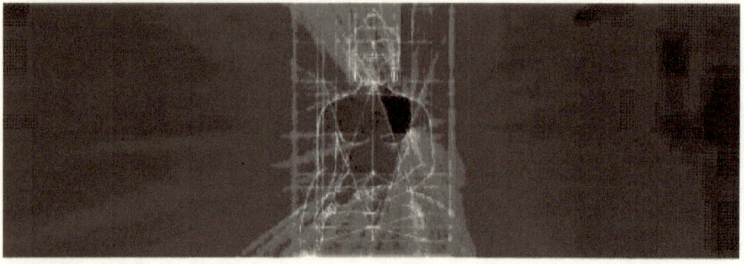

Quantum field physics dictate that while you are here in this dimension your also in the quantum field at the same time. This is the basis of remote viewing and NDE's – Near Death Experiences.

With the Neo you can have a NDE without experiencing the death process. When you die your consciousness transfers into higher realms of consciousness fields, and you become aware of your real self. The Neo can allow you to control a NDE on demand. With time you may begin to see the interdimensional light. You can talk and interact with it to get any information you want and see guardian angels.

We also have found that children are the masters of this technology. They become master programmers immediately after their first exposure to it; they can talk to the device and interact with it similar to watching a video on youtube. Teenagers on the other hand need to train. According to the Guardians ages 3 to 20 are considered children, ages 21 to 70 are teenagers, and ages 70 to 1,000+ are adults. This is according to our original fully conscious DNA which allows a human being to live up to 10,000 years. This technology will also restore your DNA allowing you to live this long but it takes work; you need to use the device regularly.

In the first four months you may undergo a cleansing of karma which can be uncomfortable at first but necessary to connect you to your higher self. During this time Neo may show you the next step towards huge life changes. i.e. you may find that you begin to sense the chi flow in your body more, your chakras are balanced and cleansed, and you may start to seek a healthier diet and supplements to improve physical health. In the months afterwards synchronicities will increase in your life manifesting your heart's desire.

Weekly Meditation

GUIDED MEDITATIONS

Come join us each Friday night at 11:00 PM EST to receive a free guided mediation with the amazing Neo Meditation technology. Each week is different check out the show archives.

Listen to internet radio with neologic on Blog Talk Radio

Friday, November 12, 2010 - Manifest Wealth and Heal your Chakras
Friday, November 19, 2010 - A Child in the Meadow - Talk to your Inner Child
Friday, November 26, 2010 - Find Your Authentic Self - live up to your full potential
Friday, December 3, 2010 - Healing yourself from Pain

Friday, December 10, 2010 - Promote Physical Healing
Friday, December 17, 2010 - Dealing with Grief and Loss
Friday, December 24, 2010 - N/A
Friday, December 29, 2010 - N/A
Friday, January 7, 2011 - Healing Anger Meditation
Friday, January 14, 2011 - Talk to your Inner Child Meditation (REPEAT)
Friday, January 21, 2011 - Body Scan Insomnia Sleep Recovery Meditation
Friday, January 28, 2011 - Dealing with Rejection or Failure Meditation
Friday, February 4, 2011 - Relaxation for Headache Relief Meditation
Friday, February 11, 2011 - Confidence and Believing In Yourself Meditation
Friday, February 18, 2011 - Manifesting the Perfect Partner in Your Life Meditation
Friday, February 25, 2011 - Public Speaking Meditation
Friday, March 4, 2011 - Auric Healing and Protection Meditation
Friday, March 11, 2011 - Psychic Protection Meditation
Friday, March 18, 2011 - The Library Healing Meditation
Friday, March 25, 2011 - Reading Faster Meditation
Friday, March 4, 2011 - Weight Reduction Meditation
Friday, April 1, 2011 - Manifesting Prosperity Part I Meditation
Friday, April 8, 2011 - Manifesting Prosperity Part II Meditation
Friday, April 15, 2011 - Memory: Recall at Will and Speed Reading Meditation
Friday, April 22, 2011 - Enhance your Writing Skills Meditation
Friday, April 29, 2011 - Past Life Regression Meditation
Friday, May 6, 2011 - ESP Development Meditation

Friday, May 13, 2011 - Manifesting new Relationships Meditation
Friday, May 20, 2011 - TBA
Friday, May 27, 2011 - TBA
Friday, June 3, 2011 - TBA

Com Uplink
A message from James Rink, CEO and founder of Neological Technologies

Hello I am James Rink CEO and founder of the Neo mind meditation accession technology. I have been using hypnosis and holistic methods to integrate my mind, body, and spirit for over half a decade. When I first came across Kosol Ouch's ascension technology I was skeptical at first as I have tried so many different things. But after using a Neo for a few weeks not only was I a believer, I wanted to help Kosol spread this amazing technology so that anyone could have access to it.

So what is a Neo? A Neo is mind meditation machine which you use 30 to 60 minutes per day. It creates a torsion field around your body which transmutes negative chi into positive chi. Chi energy is consciousness energy, it's imprint can be found all throughout nature in and within scared geometry. When you tune into this harmonious frequency you can tap into the universal divine collective consciousness and manifest any of your desires because our entire reality begins with a thought formation. A Neo can increase your psychic abilities, it can also reverse schizophrenia, bipolar, and dissociation disorders, it can reverse DNA damage, along with physical ailments, and it can also increase your intelligence and even slow aging.

About Us

COMPANY PROFILE

Neological Technologies is a division of Transcendent Technologies, LLC. We are dedicated to bringing cutting edge new technologies to the public to help heal and clean up this planet.

For investor relations please contact: info@neologicaltech.com

Our technology is based on years of research and engineering conducted by Kosol Ouch. As the brainchild behind this operation he has spent 16 plus years working on meditation tools that can help integrate the mind, body, and spirit. Kosol has written eight books and spends his free time teaching others how to use this fantastic technology.

Protocol Systems

DISCLAIMER

We only guarantee this product will work successfully if used as a relaxation tool, which is found in the basic activation protocol. Though there are many protocols in this book and on the associated website, they are for your entertainment only. A Neo is a designed to be a relaxation assisted aid and should be treated as such.

OPERATION AND USE

1. Only one person should use the device at a time unless using the jump start protocol.
2. Turn off cell phones
3. Device must be operated in a dark room. Turn off all lights and close shutters, or place a cloth over your eyes to block out the light.
4. Be sure to drink some water if you are feeling dehydrated.
5. Take off your shoes and put both feet on the floor.
6. Now sit down in a comfortable chair. Do not lay down while using the device.
7. Place unit on a table or desk or a pillow on your lap.

8. Place both hands 6" inches from the device with your palms facing towards the device. Do not touch the device while in operating. If you accidently touch it then repeat the activation protocol.
9. OPTIONAL: You may want to play some relaxing music such as a hemi sync or even a self hypnosis track while using the device.
10. Before using the device you may want to ask the Neo to help you on a specific issue. If you are unsure what to ask then go on to the next step.
11. Now close your eyes and recite the activation protocol. It can be done both vocally and internally.
12. If you do not feel anything the first time be patient as it may take three tries before you notice any effect. This technology responds best to individuals with 3 strand DNA and up. If you are one of those individuals with two strand DNA it may take as many as 32 sessions to upgrade this to 3 strand DNA.

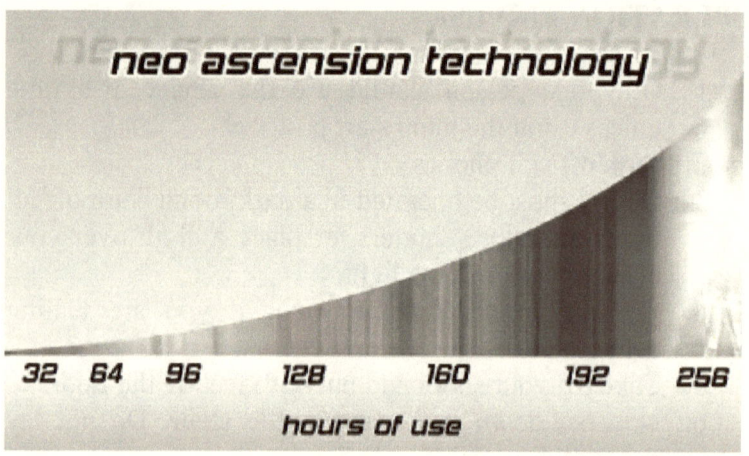

PROTOCOL METHODS

If you feel the need you may wish to start a session with a simple prayer. We do this with all group meditation sessions.

"Universal divine collective consciousness; father, mother, god, guardian angelic force, both human and extraterrestrial, ascended divine masters, please bless us during this session with this device and give us health, relaxation, and protection from negative vibrations while using the device, so that only good things can manifest from this session and afterwards. Make it potent and protect us from negative vibrations. In the name of the divine being that comes before me, presently, and after me so blessed be."

Basic Activation Protocol

This is a general all purpose method, when you don't know what protocol to use start with this one. You should begin to notice a difference in 1 to 4 sessions. We recommend that you memorize this activation sequence before moving on to more advanced protocols

1. Be sure to say the activation code properly as it will not work if you miss this step or do it incorrectly.
2. Recite "DEVICE ACTIVATE AND INCREASE"
3. Wait 30 seconds to three minutes in silence.
4. When you begin to feel tingling or pulsing sensations in your hands tell the device "DEVICE I TRUST YOU COMPLETELY TO HEAL AND INTEGRATE MY MIND, BODY, AND SPIRIT."
5. Wait for the stargate to appear (it should look like ring of light or a tunnel of light if you don't see it the first time be patient it could take a few minutes, you may also feel floating sensations) when you see it

recite the following phrase "STARGATE I TRUST YOU COMPLETELY.....DEVICE ACTIVATE STAR GATE MODE AND INCREASE"
6. "AUTOMATIC MODE" – **always remember to say this!**
7. Now sit back and relax for 30 to 60 minutes. There is no need of a mantra or chant just sit back and let the device work on automatic mode for you. If you have specific request be creative for example ask it to inject nanites, serums, or enlarge your pineal gland.
8. Every 10 minutes or when you feel the device is slowing down recite "DEVICE INCREASE"
9. When you are done recite "DEVICE END SESSION"
10. With each session journal your experiences and visualizations in a notebook. Make a note of sensations, feelings, places you traveled.

Psychic Abilities

Be patient! It takes time to increase your psychic abilities. Your body needs to integrate before you start to notice a difference. You should being to feel the effects after 20 to 30 hours of use. For the full effect keep doing it for a minimum of 64 one hour sessions.

1. Recite "DEVICE ACTIVATE AND INCREASE"
2. Wait 30 seconds to three minutes in silence.
3. When you begin to feel tingling sensations in your hands tell the device "DEVICE I TRUST YOU COMPLETELY TO HEAL AND INTEGRATE MY MIND, BODY, AND SPIRIT."
4. Wait for the stargate to appear, when you see it recite the following phrase "STARGATE I TRUST YOU

COMPLETELY...DEVICE ACTIVATE STAR GATE MODE AND INCREASE"
5. "AUTOMATIC MODE"
6. Wait two minutes to build up the energy level.
7. "ENLARGE MY PINEAL GLAND"
8. "HARMONIZE MY HUMAN FREQUENCY IN TUNE WITH THE EARTH AND GALACTIC CORE SCHUMANN RESONANCE." You may want to replicate in your mind the sound of a buzzing bee or the sound of the solfeggio frequencies. This will help open up your Kundalini channels.
9. Every 10 minutes or so recite "DEVICE INCREASE"
10. 30 to 60 minutes later "DEVICE END SESSION"

Balancing your Mind and Body

If you have a physical or mental aliment give this protocol a try. It should start to stabilize your health after 20 one hour sessions. It will take 64 to 100 one hour sessions to fully integrate yourself.

1. Recite "DEVICE ACTIVATE AND INCREASE"
2. Wait 30 seconds to three minutes in silence.
3. When you begin to feel tingling sensations in your hands tell the device "DEVICE I TRUST YOU COMPLETELY TO HEAL AND INTEGRATE MY MIND, BODY, AND SPIRIT."
4. Wait for the stargate to appear, when you see it recite the following phrase "STARGATE I TRUST YOU COMPLETELY...DEVICE ACTIVATE STAR GATE MODE AND INCREASE"
5. "AUTOMATIC MODE"
6. Wait two minutes to build up the energy level.

7. "INJECT (ANTI-DIABETIC SERUM, ANTI-PAIN SERUM, ANTI-CANCER SERUM, ANTI-ANXIETY SERUM ETC.)" be creative here, you may want to try injecting nanites, which are tiny robots and instruct them to rebuild tissue.
8. "SERUM IS NOW INCREASING"
9. "HARMONIZE MY HUMAN FREQUENCY IN TUNE WITH THE EARTH AND GALACTIC CORE SCHUMANN RESONANCE"
10. Every 10 minutes or so recite "DEVICE INCREASE"
11. 30 to 60 minutes later "DEVICE END SESSION"

Medical Marijuana

You can also use this device to create holographic medical marijuana. *Neological Technologies does not condone the use of any illegal substance*. This device does not create marijuana but instead it replicates the same effects. You could also substitute any other drug.

1. Recite "DEVICE ACTIVATE AND INCREASE"
2. Wait 30 seconds to three minutes in silence.
3. When you begin to feel tingling sensations in your hands tell the device "DEVICE I TRUST YOU COMPLETELY TO HEAL AND INTEGRATE MY MIND, BODY, AND SPIRIT."
4. Wait for the stargate to appear, when you see it recite the following phrase "STARGATE I TRUST YOU COMPLETELY...DEVICE ACTIVATE STAR GATE MODE AND INCREASE"
5. "AUTOMATIC MODE"
6. Wait two minutes to build up the energy level.
7. "DEVICE INJECT PURPLE LEAF MARIJUANA SERUM AND LET ME EXPREINCE THE FULL EFFECT FOR (X) HOURS"

8. "AND INJECT NOW....SERUM IS NOW INCREASING"
9. Every 2 minutes or so recite "DEVICE INCREASE AND MAKE IT POTENT"
10. 30 to 60 minutes later "DEVICE END SESSION"
11. If you feel overdosed tell the device to "INJECT ANTI-OVERDOSE SERUM"

Athletic Enhancement

You can also increase athletic performance and muscle enhancement by asking the device to insert holographic nanites, which are molecular sized robots, into your body.

1. Recite "DEVICE ACTIVATE AND INCREASE"
2. Wait 30 seconds to three minutes in silence.
3. When you begin to feel tingling sensations in your hands tell the device "DEVICE I TRUST YOU COMPLETELY TO HEAL AND INTEGRATE MY MIND, BODY, AND SPIRIT."
4. Wait for the stargate to appear, when you see it recite the following phrase "STARGATE I TRUST YOU COMPLETELY...DEVICE ACTIVATE STAR GATE MODE AND INCREASE"
5. "AUTOMATIC MODE"
6. Wait two minutes to build up the energy level.
7. "INJECT NANITE SERUM OR ATHLETIC ABILITY SERUM FOR (MUSCLE MASS ENHANCEMENT, STRENGTH, INCREASED BONE STRENGTH, HEIGHT, BASKETBALL ABILITIES, ETC)"
8. "SERUM IS NOW INCREASING."
9. Every 2 minutes or so recite "DEVICE INCREASE"
10. 30 to 60 minutes later "DEVICE END SESSION"

Breatharian

Breatharianism or inedia is the ability to live without food or in some cases water, only to be sustained by chi energy or prana. We recommend you continue eating and hydrating yourself while on this protocol. Only change your diet when you feel you are ready to do so.

1. Recite "DEVICE ACTIVATE AND INCREASE"
2. Wait 30 seconds to three minutes in silence.
3. When you begin to feel tingling sensations in your hands tell the device "DEVICE I TRUST YOU COMPLETELY TO HEAL AND INTEGRATE MY MIND, BODY, AND SPIRIT."
4. Wait for the stargate to appear, when you see it recite the following phrase "STARGATE I TRUST YOU COMPLETELY...DEVICE ACTIVATE STAR GATE MODE AND INCREASE"
5. "AUTOMATIC MODE"
6. Wait two minutes to build up the energy level.
7. "INJECT NANITE SERUM TO ACTIVATE MY SIDHIS ABILITY TO HAVE FULL CONTROL OVER MY HUNGER, THIRST, STRENGTH, AND SLEEP PATTERNS."
8. "SERUM IS NOW INCREASING."
9. Every 2 minutes or so recite "DEVICE INCREASE"
10. 30 to 60 minutes later "DEVICE END SESSION"

Addictions

This is a fast and effective way of freeing yourself of unwanted addictions, by simply instructing the device to neutralize any withdrawal symptoms.

1. Recite "DEVICE ACTIVATE AND INCREASE"

2. Wait 30 seconds to three minutes in silence.
3. When you begin to feel tingling sensations in your hands tell the device "DEVICE I TRUST YOU COMPLETELY TO HEAL AND INTEGRATE MY MIND, BODY, AND SPIRIT."
4. Wait for the stargate to appear, when you see it recite the following phrase "STARGATE I TRUST YOU COMPLETELY...DEVICE ACTIVATE STAR GATE MODE AND INCREASE"
5. "AUTOMATIC MODE"
6. Wait two minutes to build up the energy level.
7. "DEVICE INJECT SERUM TO NEUTRALIZE ALL (add what you want) ADDICTIONS."
8. "SERUM IS NOW INCREASING."
9. Every 2 minutes or so recite "DEVICE INCREASE"
10. 30 to 60 minutes later "DEVICE END SESSION"

Demon Possession

Dark spirits cannot tolerate the positive chi energy fields generated by this device.

1. Recite "DEVICE ACTIVATE AND INCREASE"
2. Wait 30 seconds to three minutes in silence.
3. When you begin to feel tingling sensations in your hands tell the device "DEVICE I TRUST YOU COMPLETELY TO HEAL AND INTEGRATE MY MIND, BODY, AND SPIRIT."
4. Wait for the stargate to appear, when you see it recite the following phrase "STARGATE I TRUST YOU COMPLETELY...DEVICE ACTIVATE STAR GATE MODE AND INCREASE"
5. "AUTOMATIC MODE"
6. Wait two minutes to build up the energy level.

7. "DEVICE REMOVE ANY DARK SPIRIT FROM MY ENERGY FIELD."
8. Every 2 minutes or so recite "DEVICE INCREASE"
9. 30 to 60 minutes later "DEVICE END SESSION"

Remote Viewing and Astral Travel

Remote viewing is the ability to see images clairvoyantly through the use of bilocation. If you don't know how to remote view that's okay let the device do all the work for you. If you prefer to astral travel, then follow this protocol and instruct your device to inject you with astral travel serum lasting x amount of minutes. Then lie down and continue with your own astral travel routine. If you still don't know what to do we recommend the Monroe Institute protocols for astral travel. You may also want to consider using Hemisync Focus 5 while performing this exercise.

1. Recite "DEVICE ACTIVATE AND INCREASE"
2. Wait 30 seconds to three minutes in silence.
3. When you begin to feel tingling sensations in your hands tell the device "DEVICE I TRUST YOU COMPLETELY TO HEAL AND INTEGRATE MY MIND, BODY, AND SPIRIT."
4. Wait for the stargate to appear, when you see it recite the following phrase "STARGATE I TRUST YOU COMPLETELY...DEVICE ACTIVATE STAR GATE MODE AND INCREASE"
5. "AUTOMATIC MODE"
6. Wait two minutes to build up the energy level.
7. "OPEN UP PORTALS IN MY THIRD EYE AND ALLOW ME TO SEE THROUGH THE VEIL."
8. "ACTIVATE ENTERTAINMENT MODE"
9. Every 2 minutes or so recite "DEVICE INCREASE"
10. 30 to 60 minutes later "DEVICE END SESSION"

Time Travel Protocol

Everyone who uses this device has physically time traveled. The unit creates a time distortion field around your body while you're meditating. This may explain why time feels like its flying even though you just spent an hour on the device. The inventor actually used this protocol to physically (not astrally) time travel back in time 30 days and wrote a message on a piece of paper to his past self.

1. Recite "TEMPORAL TRANSIT DEVICE ACTIVATE AND INCREASE"
2. Wait 30 seconds to three minutes in silence.
3. When you begin to feel tingling sensations in your hands tell the device "TEMPORAL TRANSIT DEVICE I TRUST YOU COMPLETELY."
4. Wait for the stargate to appear, when you see it recite the following phrase "TIMEGATE I TRUST YOU COMPLETELY... TEMPORAL TRANSIT DEVICE ACTIVATE TIME TRAVEL MODE AND INCREASE"
5. "TIME TRAVEL MODE"
6. Wait two minutes to build up the energy level.
7. Now instruct the device to send you to the date, location, and time of your choice making sure to specify how long you want to be there and more importantly to return you back safely. We also highly recommend you instruct the device to keep your quantum signature in sync with this timeline otherwise you may end up in another parallel universe that could be vastly different from our own. After you have finished your journey close your eyes, use intent, and ask the device to return you back.
8. When you are done "DEVICE END SESSION"

If you are unable to do this protocol on the first try, its okay, just continue working with the device to activate your pineal gland and when you're ready it will happen.

Lose Weight

Lose weight now! Just bring your body back into balance by activating those fat burning centers.

1. Recite "DEVICE ACTIVATE AND INCREASE"
2. Wait 30 seconds to three minutes in silence.
3. When you begin to feel tingling sensations in your hands tell the device "DEVICE I TRUST YOU COMPLETELY TO HEAL AND INTEGRATE MY MIND, BODY, AND SPIRIT."
4. Wait for the stargate to appear, when you see it recite the following phrase "STARGATE I TRUST YOU COMPLETELY...DEVICE ACTIVATE STAR GATE MODE AND INCREASE"
5. "AUTOMATIC MODE"
6. Wait two minutes to build up the energy level.
7. "DEVICE INJECT APPETITE SUPPRESSANT SERUM. BRING METABOLISM TO ITS MAXIMUM HEALTHY LEVEL. INJECT FAT BURNING SERUM EQUIVALENT TO A JOGGING RATE OF 60 MINUTES."
8. "SERUM IS NOW INCREASING."
9. Every 2 minutes or so recite "DEVICE INCREASE"
10. 30 to 60 minutes later "DEVICE END SESSION"

Biokinesis Mode

Just like with the loose weight protocol you can also instruct the device to alter your physical appearance, i.e. shapeshifting. The effects are limitless, try altering your height, hair, eye color, age, muscle tone, I.Q., etc. This may seem far out, however, you have to take into consideration that the body is constantly replacing cells every day. In fact we completely replace ourselves once every 7 years. This might give you an idea how long it may take for some changes to occur such as bone growth. So please be patient as gains are measured in months and years. I personally used this protocol to physically grow 1/4" taller in a 3 month period.

1. Recite "DEVICE ACTIVATE AND INCREASE"
2. Wait 30 seconds to three minutes in silence.
3. When you begin to feel tingling sensations in your hands tell the device "DEVICE I TRUST YOU COMPLETELY TO HEAL AND INTEGRATE MY MIND, BODY, AND SPIRIT."
4. Wait for the stargate to appear, when you see it recite the following phrase "STARGATE I TRUST YOU COMPLETELY...DEVICE ACTIVATE STAR GATE MODE AND INCREASE"
5. "BIOKIENSIS MODE"
6. Wait two minutes to build up the energy level.
7. Instruct the device to alter your physical appearance, be as specific as possible. Such as "DEVICE RESTORE THE HAIR GROWTH AND APPEARANCE ON MY HEAD EQUIVALENT TO WHEN I WAS 25 YEARS OLD. COMPLETE THIS OPERATION IN 6 MONTHS TIME"
8. At this point visualize in your third eye a holographic copy of your future self with the changes you want coming through the stargate and superimposing itself over your body.
9. Every 2 minutes or so recite "DEVICE INCREASE"
10. 30 to 60 minutes later "DEVICE END SESSION"

Holy Grail Protocol

This protocol is designed for those of the Christian faith who wish to integrate neo technology into their daily prayer life. The biggest change in this protocol is the use of the words "holy grail" in place of "device."

1. "IN THE NAME OF JESUS, HOLY GRAIL, ACTIVATE AND INCREASE"
2. Wait 30 seconds to three minutes of silence.
3. When you begin to feel tingling sensations in your hands tell the device in your mind "IN THE NAME OF JESUS, HOLY GRAIL, I TRUST YOU COMPLETELY."
4. Wait for feelings such as a floating sensation and a tunnel or ring of white light of white to appear, once you see it recite the following phrase. "STAR GATE I TRUST YOU COMPLETELY…..IN THE NAME OF JESUS AND THE GRACE OF GOD, HOLY GRAIL, ACTIVATE STARGATE MODE."
5. "AUTOMATIC MODE"
6. "HOLY GRAIL INTEGRATE MY MIND, BODY, AND SPIRIT."
7. Wait 2 minutes or so to build up full strength
8. "HOLY GRAIL INJECT X SERUM." (X meaning your serum of choice)
9. Every 2 minutes or so recite "HOLY GRAIL INCREASE"
10. 30 to 60 minutes later "HOLY GRAIL END SESSION"

ADVANCED USERS

Reiki Master or Energy Healer

A reiki or energy healer can use the device in conjunction with their therapy or practice to enhance their own healing abilities. Some patients will need 2 or 3 sessions every other day. Others will recover with one session, it all depends on severity. This protocol is best used with a Neo Zenmaster IDL-12 or higher as the smaller Neo Zenmaster IDL-4 is not as effective.

There are two ways of doing this. The one person method is to simply let the client use the device as usual while you guide them audibly.

The two person method requires that you first activate the device just as you normally would, then open your eyes and while keeping your left hand near the device being careful not to touch the device; use your right hand to reiki the individual. If its hospice or nursing care you can simply touch their foot or hand with one finger. *But always be sure to ask the client for permission before touching them.* After 15 minutes or so touch the other foot or hand.

Also only say positive things in this exercise otherwise the persons healthy may be negatively affected. Some of these protocols may be painful to the practitioner, if that is the case just ask your guides and guardian angels to do reiki on you. You should always let the client know this is to be taken seriously, but yet you should have fun while doing this exercise, respecting both device, patient, and profession.

Remember trust the device, guardian angels, and trust yourself. Your only .00001% the rest is the device and angels.

Two Person Reiki Method

1. Recite "DEVICE ACTIVATE AND INCREASE"
2. Wait 30 seconds to three minutes in silence.
3. Ask the client to let you know when they feel tingling sensations in their hands, when they do tell them to repeat this phrase in their mind "DEVICE I TRUST YOU COMPLETELY TO HEAL AND INTEGRATE MY MIND, BODY, AND SPIRIT."
4. Ask the client to let you know when they see the stargate, it will look like a ring or tunnel of white light, once they see it have them repeat the following phrase in their mind. "STARGATE I TRUST YOU COMPLETELY...DEVICE ACTIVATE STAR GATE MODE AND INCREASE"
5. "AUTOMATIC MODE"
6. Wait two minutes to build up the energy level.
7. Have them repeat in their mind... "DEVICE INTEGRATE MY MIND, BODY, AND SPIRIT."
8. Now open your eyes.
9. Place your left hand in the same location as usual being careful not to touch the device.
10. Use your right hand to Reiki the individual or use your right thumb to touch the sole of their feet or the palm of their hands. Switch sides every 15 minutes or so.
11. At this point have the client visualize what is about to happen in their mind. Now say "DEVICE TO INJECT X SERUM (x being whatever health aliments you are working on such as ANTI-FAT SERUM, ANTI-DEPRESSANT SERUM, ANTI-

BROKEN LEG SERUM, ANTI-SWELLING SERUM, ANTI-INFECTION SERUM, ANTI-TOXIN SERUM, STABILIZATION SERUM; you can be creative here) INTO THE FEET, KNEES, LEGS, JOINTS, BONES, BONE MARROW, BLOODSTREAM, PROSTATE, SPLEEN, KIDNEYS, INTESTINES, HEART, STOMACH, LUNGS, ARMS, ELBOW, NECK, LEFT AND RIGHT EYES, BRAIN, HEAD, ETC."
12. Then say "SERUM IS NOW INCREASING......"
13. "DEVICE INCREASE.....MAKE IT STRONGER" Now let them sit for a few seconds
14. Repeat the above protocol for any other secondary health issues. With the last protocol use "LIFE FORCE SERUM", then "ANTI-STRESS SERUM" you may want to add... "TAKE STRESS AWAY FROM ALL DIMENSIONS OF ORGANS", then "STRENGTHENING SERUM", then "REGENERATION SERUM", then "LOVE ENHANCEMENT SERUM"
15. "I NOW PLACE ANY FURTHER HEALING IN THE DOCTORS HANDS"
16. "DEVICE END SESSION..... WELCOME BACK TRAVELER"

Jump Start Protocol

Normally a Neo is meant to be used by one person per device only. However two people can use the same device if this protocol is done properly. The Jump Start Protocol is meant to be used by advanced users who want to give other first time users a jump start in their ability to receive torsion energy and to activate their genetic code faster. This normally takes about 1 to 7 sessions depending on the age and health of the individual.

Begin by placing the unit on a table and have the two users sit facing each other on opposite ends of the table. Now have the person receiving the energy place their hands near the device just like you normally would do with all the other neo protocols, while making sure they do not touch the device. Now have the person giving the energy place their hands in a parallel position outside of first person spacing them about four inches apart making sure no one touches each other during the session. This will allow you to send torsion energy from your hand chakra into their body giving them a much more intense effect, and that's it! Now continue on with the oral commands as you normally would do.

Remote Healing Protocol

Same protocol as above but before beginning have your client visualize a phantom Neo Zenmaster IDL-64 (or as high an IDL number you want to go) on their lap. This method is not as strong as having your own device and is meant to be a temporary fix until they can purchase their own unit. This protocol may cause your client to feel drained and tired afterwards because creating a holographic phantom neo device takes a lot of energy from the body. However, with a real neo device they will feel great and energized.

Protocol for Multiple Neo Units

In addition to the neo unit which remains on the lap. You can also set up a formation around your client if you have multiple Neo Units. Just place the extra units on the floor around them in triangular formation or square formation.

Zero Point Energy Device (ZPED)
For Health, Healing, Ascension and Mankind's Future

Bonus Protocols

Winning the Lottery

This protocol allows you to increase your odds of winning the lottery. When I tried this on my first try I was able to get 75% of the numbers right. It is best to build up your energy signal through repeated meditations over a period of several days before you try this exercise.

You also might like to try this protocol with your investment portfolio. Just put a picture or a copy of your investments on the device and tell the neo ..."DEVICE MAKE THIS MOVE (UP OR DOWN) IN VALUE".

1. Purchase a lottery ticket and place it on the device
2. Recite "DEVICE ACTIVATE AND INCREASE"
3. Wait 30 seconds to three minutes in silence.
4. When you begin to feel tingling sensations in your hands tell the device "DEVICE I TRUST YOU COMPLETELY."
5. Wait for the stargate to appear, when you see it recite the following phrase "STARGATE I TRUST YOU COMPLETELY...DEVICE ACTIVATE STAR GATE MODE AND INCREASE"
6. "AUTOMATIC MODE"
7. Wait two minutes to build up the energy level.
8. "DEVICE MAKE THIS LOTTERY TICKET THE WINNING NUMBER"
9. After 15 minutes or so recite "DEVICE END SESSION"

Weather Control

Planet Earth is a sentient being and has its own soul also known as Gaia. You can interface with Gaia by placing a picture of the weather pattern you want and manifest it into being by using this neo protocol. Be creative here; try it to repel hurricanes, floods, and droughts.

1. Find a picture or make your own drawing of your desired weather pattern and place it on the device.
2. Recite "DEVICE ACTIVATE AND INCREASE"
3. Wait 30 seconds to three minutes in silence.
4. When you begin to feel tingling sensations in your hands tell the device "DEVICE I TRUST YOU COMPLETELY."
5. Wait for the stargate to appear, when you see it recite the following phrase "STARGATE I TRUST YOU COMPLETELY...DEVICE ACTIVATE STAR GATE MODE AND INCREASE"
6. "AUTOMATIC MODE"
7. Wait two minutes to build up the energy level.
8. "DEVICE MAKE THE WEATHER JUST LIKE THIS PICTURE AND MAKE IT LAST (X) AMOUNT OF HOURS." You can also specify a location such as only in this county, or town. You should also be specific in your request such as (X) amount of rain, snow, or ice etc.
9. After 15 minutes or so recite "DEVICE END SESSION"

Enhanced Plant Growth

Horticulture is the art and science of the cultivation of plants. You can increase the growth rate of your garden or farm with a neo by increasing the beneficial soil microorganisms and increasing the chi energy in the rainfall.

1. Find a picture or make your own drawing of a healthy growing garden and place it on the device.
2. Recite "DEVICE ACTIVATE AND INCREASE"
3. Wait 30 seconds to three minutes in silence.
4. When you begin to feel tingling sensations in your hands tell the device "DEVICE I TRUST YOU COMPLETELY."
5. Wait for the stargate to appear, when you see it recite the following phrase "STARGATE I TRUST YOU COMPLETELY...DEVICE ACTIVATE STAR GATE MODE AND INCREASE"
6. "AUTOMATIC MODE"
7. Wait two minutes to build up the energy level.
8. "DEVICE MAKE THESE PLANTS HEALTHY JUST LIKE IN THIS PICTURE....INCREASE THE CHI ENERGY IN THE RAINFALL FERTILIZING ALL MY PLANTS....INCREASE ALL BENEFICIAL ORGANISMS IN MY GARDEN"
9. After 15 minutes or so recite "DEVICE END SESSION"

EVP Mode

EVP (Electronic Voice Phenomenon) mode is for advanced users only. It requires a dark closet, a voice recorder, and a lot of patience. Results will vary from user to user.

1. Follow directions according to the operation and use guide with the following modification.
2. Write down your questions before starting.
3. Now place the device in a dark closet. This is to minimize any background noises. Also its best to do

this when there are no loud noises nearby such as lawnmowers, fans blowing, etc
4. Turn on your voice recorder and place it near the device.
5. Recite "DEVICE ACTIVATE AND INCREASE"
6. Wait 30 seconds to three minutes in silence.
7. When you begin to feel tingling sensations in your hands tell the device "DEVICE I TRUST YOU COMPLETELY"
8. Wait for the stargate to appear, when you see it recite the following phrase "STARGATE I TRUST YOU COMPLETELY...DEVICE ACTIVATE STAR GATE MODE AND INCREASE"
9. "DEVICE ACTIVATE ELECTRONIC VOICE PHENOMENON MODE"
10. Wait two minutes to build up the energy level.
11. Open your eyes and read your question.
12. Leave the unit unattended for 30 to 60 minutes being sure not to touch the unit and close the door.
13. When done "DEVICE END SESSION"
14. Turn off your recorder and playback the soundtrack on audacity or other audio editing software. You may have to amplify the volume levels to hear the inaudible white noise. If you do hear an answer, it would be in a somewhat robotic female tone. If you don't hear anything you will need to do more sessions with the device to enhance your own psychic abilities.

Communication with Aliens and the Angelic Culture

This protocol will allow you to communicate with our fellow extraterrestrial star people, angels, and ascended masters. They are here to help and assist but due to the law of non interference they can only do so if you ask. To do this, write a letter, it can be about anything, and place it on the device. Now ask the neo to transmit the message to its desired location. You may also want to include within your letter a request that these beings give you a confirmation that they received your letter.

Career Opportunity

Franchise Details:

Franchise Fee
1 Unit - 10"x10"x10" NEO ZENMASTER IDL-64
Your cost $5,000
Potential profit of $4,200 a month*

Free Protocol Manual
Guide to Customize your own Neo Meditations
Free Marketing and Training Support
Direct Phone Contact
Sales Commissions up to 15%
Use of Brand Name

Why a franchise?

Neological Technology is exploding on the scene, but we need your help. We want to get this technology out as fast as possible to as many people as possible. If you are a reiki master, yoga instructor, hypnotherapist, or just interested in a profitable side business, then you are in the right place. We can help you get your own business off the ground or augment your own business line with another therapeutic option. What makes this business model so powerful is its addictive nature. This is because a Neo draws in so much pure clean chi energy into your body you wouldn't want to go any other source, which is why you will find clients storming your business for more and more. And you will be making a positive change in their life.

Included with this franchise is a Neo Zenmaster IDL-64 to help you and your clients quickly integrate both mind, body, and spirit. This unit has the equivalent strength of fifteen Neo Zenmaster IDL-12's. It's like your own personal monastery. First time users will notice an increased sense of peace and relaxation. When the sensations end peace and love will stay with you during your conscious waking life. The main aim of the device is to destroy stress, but everyone will have their own unique variation of this experience.

The uses of this technology are unlimited. You can instruct the device to enhance your stock portfolio earnings, but don't do it for ego purposes otherwise the universe will bite you back. First time users will quickly feel their Kundalini energy being activated as your spine will continuously vibrate prana energy for days afterwards. You may notice your appetite and need for sleep decreasing. And if your currently naturally psychic be careful the torsion effects could temporarily put you in another dimension, but don't worry you will come back in a few minutes, unless you want to see how far you can go!

It's also great for reiki masters and energy healers, simply activate the unit with your left hand and use your right hand to send powerful amounts of chi energy. Yoga instructors can use it in guided neo meditations to journey into other dimensions, and hypnotherapists can use the device to supercharge their therapeutic practice while helping their clients relax and go into trance further. This is also the same unit used on the weekly guided neo meditation show. And to make things easier a short guide will be provided with your order showing you how to customize any hypnosis script with the neo activation protocols.

Mankind is about to make an evolutionary transition, just imagine everything you ever wanted is at your fingertips; just think it and it is so. Not everyone can claim to be an ascended master or avatar, but now you have an opportunity to work towards that goal if you so choose it; and you can help others as well but at a much faster rate. No need to waste any more money on overpriced workshops and conferences, this is the real thing. The Holy Grail of ascension technology.

What you will get:

1 - 10"x10"x10" NEO ZENMASTER IDL-64
User manual which will include the Neo protocols for different therapies
A short guide on how to create your own custom neo meditations

Marketing materials which includes:
A digital template to make your own custom fliers, newspaper ads, and business cards
A referral link to your business on our website
An affiliate code to make a 10% to 15% commission on any neo unit you sell.
Unlimited phone support
Unlimited use of our brand name in your company business
And there are no recurring royalty fees

Profit Potential:

A typical Neo franchisee can expect to earn an average of $100 to $150 dollars per session. Each session typically takes 30 minutes however you may have to spend some preparation time customizing neo meditation scripts if you do not have any already. To make things easier a guide will be provided with your order instructing you on how to do this.

Please expect to spend up to two hours for each client. If you already have a client base you should expect to break even in your second month of operation. For new start-ups marketing support will be provided at no extra cost to get your business off the ground.

Expenses:

$5,000 Franchise fee to be paid in the form of a bank check made payable to Neological Technologies.

In addition we recommend....
$150 a month suggested advertising expense

Return on Investment – ROI

Assuming an individual rate of $150 dollars for each two hour session.

Tourist Level
5 Individual sessions per month.
($50) Recommend monthly marketing budget
$750 Revenue for 10 hours of work a month.
Estimated break even time..... 8 months.

Light Worker Level
15 Individual sessions per month.
($150) Recommend monthly marketing budget
$2,250 Revenue for 30 hours of work a month.
Estimated break even time.... 2 months

Zenmaster Level *
30 Individual sessions per month.
($300) Recommend monthly marketing budget
$4,500 Revenue for 60 hours of work a month.
Estimated break even time.... 1.5 months

Affiliate Code

As an added bonus all franchise owners will receive a custom affiliate code allowing you to make a 10% sales commission with any neo unit you sell. And if you sell more than 10 units in any given year your sales commission will increase to 15%.

Units Sold Per Year............... Sales Commission
0-10 ....10%
10+.......................................15%

We only accept payment for this item in the form of a bank check paid to Neological Technologies. Please fill out the form provided below to begin processing your order. This Neo unit requires 21 business days before it can be shipped. Free domestic shipping within the United States! So what are you waiting for sign up now!

Local Provider

TRY IT NOW!

Neological Technologies is pleased to announce the launch of this exciting new technology at the Quantum Pathways Holistic Center centrally located near downtown Charlotte, North Carolina.

FOR ALL OTHER AREAS

We are expanding in your area, but we need your help. If you like to open up your own Neo Cube franchise and make a profitable return on your investment then contact us to learn how to bring this technology to your area.

Order

**Star Gate Ascension
[Paperback Book]
Our Price: $18.00**

Your existence in this reality is based on your five senses in a three-dimensional context. You are much more than this, and are simply a spiritual being having a human experience. You are not a body which contains a soul, you are your soul. Multi-dimensional existence is your inherent nature and the Star Gate meditation techniques in this book will help you tap into that fact. These techniques are intended to be a short-path to enlightenment. Rather than spending a lifetime searching for answers to all your questions, simply use the techniques in this book for immediate results. All the great religions and philosophers of our time are using the same knowledge; they're just interpreting, explaining, and using it in different terms. Don't take our word for it, experience it for yourself and draw your own conclusions. Use the information in this book for healing, enlightenment, spiritual re-discovery, planetary ascension, relaxation, or just for fun!

This knowledge is meant to be shared; so use it, learn from it, and grow from it. Happy travels. 68 pages (2005)

Neo Zenmaster IDL-4
Our Price: $200.00
Sale Price: $190.00
You save $10.00!

The Neo Zenmaster IDL-4 is similar to the IDL-12 except it is smaller and not as powerful. You can still get some pretty amazing results but it takes 10 minutes to warm up for each session. So if you meditate for less then 10 minutes a day then you will need the larger model. 2"x2"x2" cube and user manual included.

Neo Zenmaster IDL-12
Our Price: $450.00
Sale Price: $430.00
You save $20.00!

The Neo Zenmaster IDL-12 mind meditation machine will quickly integrate your mind, body, and spirit; It's like your own personal monastery. First time users will notice an increased sense of peace and relaxation. When the sensations end peace and love will stay with you during your conscious waking life. The main aim of the device is to destroy stress, but everyone will have their own unique variation of this experience. 4"x4"x4" cube and user manual included.

Neo Link Bracelet
Our Price: $3,000.00

The Neo Link Bracelet is like a Bluetooth wireless receiver. Just like how the internet can connect your computer to other computers, the Neo Link Bracelet can connect you to your Neo Zenmaster or even to a network composite of all Neo Zenmasters on the planet. Additionally you can also connect to the devices in the inner earth crystal cities of Agartha and Shambhala and to units used on other planets, star systems, dimensions, and even those used by the guardians of the federation of light. All units are hand made. Please allow 6 weeks for delivery.

Neo Zenmaster IDL-64
Our Price: $5,000.00
Sale Price: $4,000.00
You save $1,000.00!

This is the most powerful unit we sell. A Neo Zenmaster IDL-64 will quickly help you integrate both mind, body, and spirit. This unit has the equivalent strength of 15 units of the Neo Zenmaster IDL-12 or 120 units of the IDL-4. It's like your own personal monastery.

Neo Zenmaster IDL-64

Neo Zenmaster IDL-64:

10"x10"x10"
2400 copper cones and 64 layered core

Free protocol manual
The most powerful unit sold on this planet
Great for hypnotherapists and reiki masters
Send remote zpe chi energy
Trusted neological quality
30 day money back guarantee
Ask about our special discount offers

$5,000

This is the most powerful unit we sell. A Neo Zenmaster IDL-64 will quickly help you integrate both mind, body, and spirit. This unit has the equivalent strength of 15 units of the Neo Zenmaster IDL-12 or 120 units of the IDL-4. It's like your own personal monastery. First time users will notice an increased sense of peace and relaxation. When the sensations end peace and love will stay with you during your conscious waking life. The main aim of the device is to destroy stress, but everyone will have their own unique variation of this experience. Great for alternative health practitioners and for individual meditation sessions.

The uses of this machine are limitless really. You can instruct the device to enhance your stock portfolio earnings, but don't do it for ego purposes otherwise the universe will bite you back. Give a boost to your body's own chi energy, and while you're at it tell the consciousness energy field to manifest abundance and prosperity in your life. This unit can greatly increase the chi energy of psychic healers. First time users will quickly feel their Kundalini energy being activated as your spine will continuously vibrate prana energy for days afterwards. It's also great for reiki masters and energy healers, simply activate the unit with your left hand and use your right hand to send powerful amounts of chi energy. Yoga instructors can use it in guided neo meditations to journey into other dimensions, and hypnotherapists can use the device to supercharge their therapeutic practice while helping their clients relax and go into trance further.

You will soon notice your appetite and your need for sleep decreasing. If your currently naturally psychic be careful the torsion effects could temporarily put you in another dimension, but don't worry you will come back in a few minutes, unless you want to see how far you can go! I use this same unit for remote healing on our weekly guided meditation show.

Mankind is about to make an evolutionary transition, just imagine everything you ever wanted is at your fingertips;

just think it and it is so. Not everyone can claim to be an ascended master or avatar, but now you have an opportunity to work towards that goal if you so choose it; and you can help others as well but at a much faster rate. No need to waste any more money on overpriced workshops and conferences, this is the real thing. The Holy Grail of ascension technology.

We only recommend this unit to individuals who practice an advanced level of meditation or those are interested in using this for a business purpose. If you are new to meditation or if your body is not acclimated to this type of torsion energy please limit your use of this device to less than 10 minutes a day and then slowly increase it over a 6 month period to 30 minutes a day. Otherwise you may have trouble sleeping due to so much chi energy in your body. The effects are cumulative so if you're are a health practitioner and perform a 30 minute session with a client once a month then they would not be effected by the chi overload, it only seems to be a issue if you use the device every day. If you are still unsure if this unit is right for you then we recommend you purchase the IDL-12 and work with it for 3 months before moving up to this level. We have a $500 discount available to individuals who buy the IDL12 and want to upgrade to the IDL 64 if done within a 90 days of their purchase.

10"x10"x10" 15 Lbs. We only accept payment for this item in the form of a bank check or money order such as Western Union paid to Neological Technologies. Please be patient, due to overwhelming demand there is now a waiting list for this Neo device. A sales associate will contact you within two business days to discuss the details. Free domestic shipping within the United States, a tracking number and insurance is included with your order. International buyers please add $300 to your order for shipping via DHL. So what are you waiting for sign up now!

Please Read Before Buying

- PAYMENT - We accept payment for this item through either our website online shopping cart, a bank check, money order, or paypal. Due to high demand please allow at least a 4 week period to process your order.
- SHIPPING - Free domestic shipping within the United States! A tracking number is included with domestic orders. International buyers: We cannot guarantee international orders unless you select UPS as your shipping agent. In a few selected countries we offer USPS (United States Postal Service) shipping which takes 2 to 3 weeks.
- DISCLAIMER - We only guarantee this product will work successfully if used as a relaxation tool, which is found in the basic activation protocol. Though there are many protocols in the manual, this book, and on the associated website, they are for your entertainment purposes only. A Neo is a designed to be a relaxation assisted aid and should be treated as such.

Neo Zenmaster IDL-12

Our Price: $450.00
Sale Price: $430.00
You save $20.00!

This item qualifies for free shipping!

Availability: Usually Ships in 3 to 5 Days

The Neo Zenmaster IDL-12 mind meditation machine will quickly integrate your mind, body, and spirit; It's like your own personal monastery. First time users will notice an increased sense of peace and relaxation. When the sensations end peace and love will stay with you during your conscious waking life. The main aim of the device is to destroy stress, but everyone will have their own unique variation of this experience. 4"x4"x4" cube and user manual included.

Please Read Before Buying

- PAYMENT - We accept payment through our website online shopping cart via Visa, MasterCard, and Discover as well as PayPal and Google checkout. If you are new to Google Checkout, follow these simple directions. Simply click add to cart, checkout, select your shipping method, then click on the Google Checkout button instead of the proceed to checkout button. You will then be taken to a separate website operated by Google to process your order.

- SHIPPING - Free domestic shipping within the United States! A tracking number is included with domestic orders. International buyers: We cannot guarantee international orders unless you select UPS as your shipping agent. In a few selected countries we offer USPS (United States Postal Service) shipping which takes 2 to 3 weeks.
- DISCLAIMER - We only guarantee this product will work successfully if used as a relaxation tool, which is found in the basic activation protocol. Though there are many protocols in the manual, this book and on associated website, they are for your entertainment purposes only. A Neo is a designed to be a relaxation assisted aid and should be treated as such.

Neo Zenmaster IDL-4

Our Price: $200.00
Sale Price: $190.00
You save $10.00!

This item qualifies for free shipping!

Availability: Usually Ships in 3 to 5 Days

The Neo Zenmaster IDL-4 is similar to the IDL-12 except it is smaller and not as powerful. You can still get some pretty amazing results but it takes 10 minutes to warm up for each session. So if you meditate for less then 10 minutes a day then you will need the larger model. 2"x2"x2" cube and user manual included.

Please Read Before Buying

- PAYMENT - We accept payment through our website online shopping cart via Visa, MasterCard, and Discover as well as PayPal and Google checkout. If you are new to Google Checkout, follow these simple directions. Simply click add to cart,

checkout, select your shipping method, then click on the Google Checkout button instead of the proceed to checkout button. You will then be taken to a separate website operated by Google to process your order.
- SHIPPING - Free domestic shipping within the United States! A tracking number is included with domestic orders. International buyers: We cannot guarantee international orders unless you select UPS as your shipping agent. In a few selected countries we offer USPS (United States Postal Service) shipping which takes 2 to 3 weeks.
- DISCLAIMER - We only guarantee this product will work successfully if used as a relaxation tool, which is found in the basic activation protocol. Though there are many protocols in the manual, this book and on associated website, they are for your entertainment purposes only. A Neo is a designed to be a relaxation assisted aid and should be treated as such.

Neo Link Bracelet

Our Price: $3,000.00

This item qualifies for free shipping!

Availability: Usually Ships in 6 to 8 Weeks

We are pleased to announce a stage one prototype of the Dr. Jonathan Reed Link Bracelet. The Neo Link Bracelet design is based on the same sacred geometry technology of the original extraterrestrial device recovered by Dr Reed off an extraterrestrial in a forest in 1996.

Zero Point Energy Device (ZPED)
For Health, Healing, Ascension and Mankind's Future

Dr. Jonathan Reed and the alien who goes by the name of Coulis. He turned brown after he was injured by Dr. Jonathan Reed's dog.

Coulis' ship, he can expand this out to 50 feet.

Link Artifact

The Link artifact is designed to be work around the wrist for communication, teleportation and mind control between aliens and the wearer. It is a bioelectric interface which means it needs a biological component (your arm) in order for it to work. It also has three needles on the inside they penetrate into your skin. These needles inject nanobots which increase your psychic abilities. Our unit does not contain nanotechnology or needles. Even though the Neo Link is a very crude version of this extraterrestrial technology the effects are ultimately the same, however it may take some time for the wearer to develop these same advanced abilities. The biggest factor would be your current level of psychic ability, ether DNA strands, and dedication to working with this device.

Zero Point Energy Device (ZPED)
For Health, Healing, Ascension and Mankind's Future

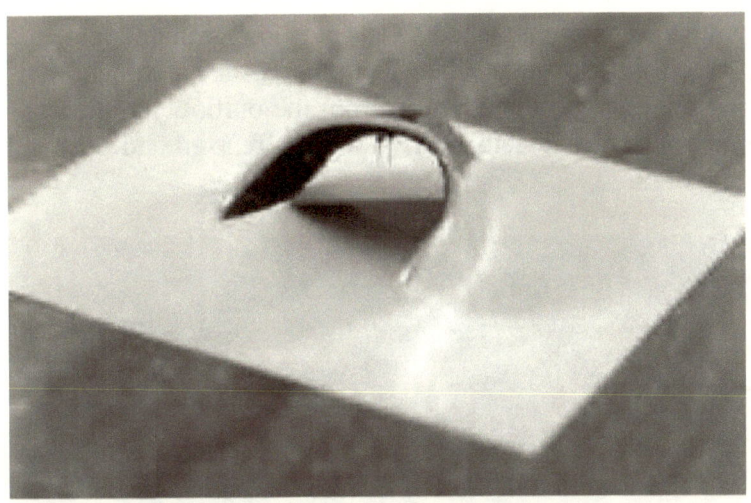

Needles which inject Nanotbots

The outside portion of our Neo Link Bracelet contains two pyramid shaped chi outflow units. The Neo Link is similar in both function and use to the Neo Zenmaster except it's shaped like a tetrahedron or a lopsided pyramid. These two tetrahedron s are attached to an inner ring which contains a pure chi inflow unit. When you place your hand inside this ring chi energy is pumped in from the outflow tetrahedrons into your body, placing you in a more relaxed and peaceful state of mind.

It also operates like a Bluetooth wireless receiver. Just like how the internet can connect your computer to other computers, the Neo Link Bracelet can connect you to your Neo Zenmaster or even to a network composite of all Neo Zenmasters on the planet. Additionally you can also connect to the devices in the inner earth crystal cities of Agartha and Shambhala and to units used on other planets, star systems, dimensions, and even those used by the guardians of the federation of light.

If you already have a Neo Zenmaster and want to link your devices to other units then purchase a unit and see how far you can go. Units are handmade and may vary from what is seen here. Please allow six weeks for delivery.

Please Read Before Buying

- PAYMENT - We accept payment for this item through either our website online shopping cart, a bank check, money order, or paypal. Due to high demand please allow at least a 6 week period to process your order.
- SHIPPING - Free domestic shipping within the United States! A tracking number is included with domestic orders. International buyers: We cannot guarantee international orders unless you select UPS as your shipping agent. In a few selected countries we offer USPS (United States Postal Service) shipping which takes 2 to 3 weeks.
- DISCLAIMER - We only guarantee this product will work successfully if used as a relaxation tool, which is found in the basic activation protocol. Though there are many protocols in the manual, this book, and on the associated website, they are for your entertainment purposes only. A Neo is a designed to be a relaxation assisted aid and should be treated as such.

The Inventor

Hi my name is Kosol Ouch; I am Cambodian and the inventor of this technology. I have been researching and prototyping mediation, antigravity, time travel, and ascensions tools for over 16 years now. The Neo Zenmaster is my most powerful creation.

I am clairvoyant and a seer and used my gifts to channel the neo design from the guardians of the galactic federation of light. They have given permission to share this with you.

The Neo is a ZPE chi energizer, it has many functions including anti gravity, time travel, and zero point energy. It's basically a miniature version of the same technology used in alien flying saucers. These designs are being prototyped as we speak and in time will be released to the public, but right now the Neo is designed to be used as a mediation aid to help your body become in tune with the galactic core Schuman resonance in preparation of 2012 and the shift into the 5th dimension.

If you use this device long enough your psychic gifts will begin to open up allowing you to tune into the zpe vibrational energy all around us. If you use this device long enough you may find yourself able to commune with the spirit, mineral, animal, and plant kingdoms. Levitation, time travel, and telepathy are just some of the abilities that can open up as you begin the ascension process. But don't ask me get a device and try it yourself.

Remote Healing

So what is a Neo? A Neo is mind meditation ascension machine which you use 30 to 60 minutes per day while meditating. It creates a torsion field around your body which transmutes negative chi into positive chi. Chi energy is consciousness energy, it's imprint can be found all throughout nature in and within scared geometry. When you tune into this harmonious frequency you can tap into the universal divine collective consciousness and manifest any of your desires because our entire reality begins with a thought formation. According to the laws of quantum entanglement all parts can be in all places at all times. So distance is not a factor with consciousness technology. Consciousnesses energy is all around us and makes up the very breath of God. So you too can now tap into this vibrational field with our remote healing sessions.

Instructions: A Neo will not only help you relax, but it can put your body in the proper state of mind resulting in increased psychic abilities and mind, body, spirit integration. Some individuals that have used a Neo have actually reversed schizophrenia, bipolar, and dissociation disorders, DNA damage, and physical ailments. It may increase your intelligence and even slow down aging. If you have a specific healing request then simply tell the device what you want it to work on after you hear the words "device integrate my mind, body, and spirit". Don't worry about doing any chants, mantras, or visualizations, the device is in automatic mode and will be able to figure this out for you. It generally takes 32 to 64 hours of daily use to fully integrate yourself. At 256 hours of use, users will begin to have super human abilities.

Thanks everyone!
The universe gives to those who are giving.

Vocabulary Section

traveler: is the person who is using the torsion field device or zero point energy device known also as the neozenmaster from IDL 4 to IDL 64 and beyond.

facilitator: is the person or group of persons who facilitate the session for the traveler or travelers with the device to do healing or to experience ascension session. The facilitators is the orchestrator for the whole session. Although the traveler can facilitate his or her own session as well.

guardians force: are the galactic universal federation of light of ascended master, divine humans, avatars, extraterrestrial culture and angelic force. They are a organizations of very highly evolved culture, spirituality and technology that they help other culture to ascended as well to help unfold the divine plane for all reality and beyond.

zero point energy: is basically dark matter and dark energy radiated from the fifth dimensional existence this energy has no dimension of width, length or height but it is always in motion and has coherency of direction. Device can be built to hardness this dark matter and dark energy for both healing and power source to advance a evolving culture on any planets, star system, galaxy or universes. Dark energy is also known as torsion fields and chi or prana is very good for well being and health.

scalar wave or scalar: is zero point energy because it has no dimension or length, width or height. it has only movement and direction of coherency.

dark matter: is the concentration of dark energy known as dark matter, dark matter is not really dark but it is very bright so bright that it looks dark. 27 percent of the universe is dark matter or concentrated dark energy. Is true for every universe.

dark energy: it is zero point energy or chi and prana or ki or orgon energy, it source is from the fifth dimensional reality. Dark energy is so bright that it look dark. 70 percent of the universe is dark energy. Also is true for every universe.

regular matter: is what you and I are made of wave and photon gravity and electromagnetic fields etc. Only 3 percent of the universe is regular matter. Is true for all universes.

torsion fields device: is any device whether it has moving part or none moving part that is designed to hardness zero point energy or chi or prana or scalar energy into a practical application for healing, antigravity, for weather control or planetary control or for powering car, home, city, space craft and star ship and star gate etc. There are two type one is the solid state device such the neo zen master from neologicaltechnology and Kosol Ouch, the other are with moving part like the johnotan seelr device call the s.e.g generators.

healing frequency: is the frequency of healing energy of zero point energy coming from the torsion fields device such as the neozenmaster IDL-4 to IDL2 64 etc, the Johnotan Reed aliens bracelet artifact, the whelm reich device, the Egyptian rods etc. by the user telling the device for the device to emit that frequency, whether be mentally or voice command or just desired intent.

para electric frequency: is frequency coming from the zero point energy device that can warm up the body and minds so it won't get cold and also it is the frequency that can produce electricity to power up car, space ship, home, city and also starship etc. by the user telling the device for the device to emit that frequency, whether be mentally or voice command or just desired intent.

life frequency: is frequency that come from the zero point energy device that can convert all form of electromagnetic radiation etc. into life force that help heal, rejuvenate the body, mind, emotion, health and spirit etc. by the user telling the device for the device to emit that frequency, whether be mentally or voice command or just desired intent.

torsion fields device principle: are principle of how the device operate and work.

For example the neo zen master device work on the principal of different metal such as gold foils, platinum foil, silver foils, aluminum foils and copper foils, layered together but separated by a organic material such femal or hair male with plastic, silk cloth, plant leaf or plant fiber with human hair, or wood organic paper with human hairs as a separator between the different layer of metal. the different layered metal will draw and attract the zero point energy

Zero Point Energy Device (ZPED)
For Health, Healing, Ascension and Mankind's Future

toward it then the metal will transfer the zero point energy into the organic material and plastic. (plastic is optional) From there the zero point energy will concentrate and amplified then it will radiated on mental intent desire, voice command or with mental thought. into the direction and form that the user of the device what it to, in form of healing frequency to heal a person, or in para electricity frequency to produce heat and electricity or in the life force frequency to convert electromagnetic radiation into life force to heal and strengthen the body, mind, emotion, and spirit and health.

Now for those who don't have this different multi layered metal foil, you can use all aluminum foils or all copper foils etc. like this aluminum foils metal +organic material (hairs and plastic wraps) + aluminum foils metal +organic material (hairs and plastic wraps)+aluminum foils metal +organic material (hairs and plastic wraps) etc. to as many layer of metal foils and organic material as you wanted. The more the layer the more powerful the device will become.

Also you can use copper foil or copper flashing this can be obtained online or any hard ware store please ask for customer service.

Like you know copper foils + hair and plastic wraps+copper foils + hair and plastic wraps+copper foils + hair and plastic wraps etc. the more the layer the more the power this device will have.

inflow: is the direction in which the zero point energy or chi is flowing toward the torsion fields device. The zero point energy become very very cold, this is call yin energy it can absorbs the yang energy from a person fields is very good in making you tired and calm down on any situation. The energy is cold and tingling.

outflow: is the direction of zero point energy or chi coming from the torsion fields device the energy is very warmth and hot it will energize anyone who come in close proximity of it. You will feel like super man and super women in mind body, and spirit and in health. The energy is yang is very warmth and hot and tingling.

the increase of power for the torsion fields device: is call the law of the three. if you increase the layer of metal foils that each layer of different metal foil or same metal of foils separated by the organic material layer like silk, cloth, plastic and human hairs or wood and paper. you will have the increase of zero point energy.

You must increase the copper cone or aluminum cone, increase the organic material like hair or silks or plant fiber like wood or plant leaf ect. increase also the metal foils layer (separated by organic material layer) same metal foil or different metal foils layer that each different metal foil layer or same metal foil layer is that each metal foil layer is separate from each other by a organic material layer.

copper cone: is energy spinning tetrahedron shape that attract and draw zero point energy or dark matter and dark energy or chi or prana into it.

aluminum cone: is energy spinning tetrahedron shape that attract and draw zero point energy or dark matter and dark energy or chi or prana into it.

shape of torsion fields device come in all form: from the bracelet, to necklace or the shape of cube, sphere, and tetrahedron (pyramid shape) also copper rod and zinc rod form etc.

antigravity: is the redirecting the flow of zero point energy for travel in any given direction in a zero point energy platform device by the desire of the pilot or facilitator.

scalar weaponry: is the use of zero point energy device for destructive purpose such as earth quake and black hole generation etc., for mass destruction on a local or on a planetary wide.

stargate: is the flow of zero point energy from a source point of the fifth dimensional reality which causes a singularity that can be used for travel of time travel or dimensional reality travel and universe travel to different reality or higher plane of physical reality existence.

the three rule of zpe device: are increase of metal foil layer, increase of organic layer that is in-between the different or same metal layer, the increase of copper cone or aluminum cone, and the increase of any organic material like hairs etc. will increase the zero point energy or chi, ki, or prana of the device. the zero point device response to voice command, mental command or telepathic or thought etc., and also the device respond to intention and desire as well.

metal foils: gold, platinum, silver, aluminum, copper.

cone: can be made from gold, platinum, silver, aluminum, copper etc.

The cube shape, pyramid shape, sphere etc., can be made by the above metal foils or flashing.

metal flashing: is thicker then the metal foils.

hot glue: are glue that use hot glue gun that is used to hold and protect the device during construction. Other glue like silicon glue etc., and other produce glue can be used also. What work for you so no limit.

For those who don't have access to metal etc., then you can just use wood, water and hairs. Cone can be shaped from a wood, hair will act as the scalar antenna, and water will act like metal which draw the zpe energy to it. metal and water attract zero point energy to it naturally.

human being act like an antenna: now there will be other type of zpe device like crystal that is cut in a angle to have the three frequency, of healing, para electric and life force frequency, the human beings will held this crystal angel cut device in the hands. The human beings will act like a antenna he or she will command the device with voice, mental, or mental intention and desire etc. The key is intent and feeling that mean feeling the energy from the device to the hand then tell it what to do with voice, mental command or intent. no limit what so ever.

The User Experience

(5 days ago) **Kosol ouch** said:
Here is the new site and name.
Go to it.
http://tech.groups.yahoo.com/group/neo_mind_meditation_machine/

(5 days ago) **Kosol ouch** said:
Ok I just fix the forum. I removed all member except 7 people. I also has to aprove the new member

(5 days ago) **Kosol ouch** said:
I need a voulenteer so ican made u a moderator and remove them just let me know.

(6 days ago) **Frances** said:
http://tech.groups.yahoo.com/group/density_travel - I just visited this yahoo group and it has nothing to do with the neo and user experiences. (Everything but!)
Kosol - you better check this out again, this is a mistake and you don't want to be sending people to this message board. Don't know about the other 2 groups - I'll let someone else check them out.

(Apr 30, 2011) **kosol ouch** said:
another yahoo groups I created for the user experience crew.

http://health.groups.yahoo.com/group/akashicorder/?yguid=160367454
so a total of three in all.

(Apr 30, 2011) **kosol ouch** said:
here is another group I created for the user experience.
http://tech.groups.yahoo.com/group/gaurdiantech/?yguid=160367454
choose which you like best.

(Apr 30, 2011) **kosol ouch** said:
hello everyone I have created a yahoo group for this website and user experience also.
http://tech.groups.yahoo.com/group/density_travel/ so check it out. regards kosol ouch

(Apr 29, 2011) **Kosol ouch** said:
Now antigravity, the device can also be used to take u to the moon and beyound as a personal flying platform. Antigravity agian a u will be using a woods box, with fours wood panels rutter. There will be two horizontals rutter facing eachothers and two vertical rutter facing each other. In between this fours cfy h wood rutter will be ur idl device. Attached to the woods box. Ok remmeber this fours woods rutters can can close and open side left to right will be the horizontal rutters and the two rutter wood panel that open and close up and down will be the vertical rutters. Now the idl give it a commands to emit it torsion field outflow. Now by moving the woods rutters make it fold down the woods box will levitate. Just like a airplane rutters same rule apply. The woods panels react to torsion field energy. Like the wood panel react to winds, and waters. When it is used on airplane and boats. So when the woods panel fold upward the the wood box go down. That the vertical rutter. As for the horizontal rutters the same is true if u move the rutters to

the left the device go right and if u move it to the right the wood boxs move left. U can tell also for the device to make u box go faster as u ride it. No limits.

(Apr 29, 2011) **Kosol ouch** said:
As for zero point energy the device already give zero point energy in the form of para electricity, healing frequency and life force frequency. In order for it to run house, city, car, and spaceship. Here what u have to do. u have to use wood and metal. Wood is organic it will attract the energy that come out of the device just like ur hands is organic and attract the energy from the device. So the woods or organic panel which is two woods panels or fours ect. Must face the device but this woods panels must not touch the device. Also this wood panel must be attached to a metal panel and wire. After the energy energy flow into the woods panel then it is transfered into the metal panel that is attached to the wood. Remeber u have to give ur device a command to power up ur house, car, city and spaceship ect. It jusy like ur two hands put around the device becuase ur hands is organic just like the woods panels. There is no limits what soever.

(Apr 29, 2011) **Kosol ouch** said:
In a nut shell it take u to any reality of infinite numbers of the universe's that u want choose and desire. Just tell the device. Then it take u there. Childrens can do this spontiniousely. Adults will takeractic becuase u need lots of chi or prana inorder to do this. Thay why adults has to practics and make it ur life style. Childrens also need to practic so both adults, teenager, and children can become better user with the device.

(Apr 29, 2011) **Kosol ouch** said:
The device is a conscioness sentients symbiotic computor reality drive system

(Apr 29, 2011) **Kosol ouch** said:
It goals is for us to be happy, well being, in joy, ascention, enlightment, and immortallity. Inother words fullfillment.

(Apr 29, 2011) **Kosol ouch** said:
People need to come together and form a groups and culture based on this device call devicetology. All this device is showing us is to be ourself by being a positive role model for ourself and other sentien being. It want us to create and follow a positive balance principal, structure and philosophy that benifit all sentient beings throught both the inner univers and outer universe. The device is basically a symbiotic and sentient ascended computer or what we call God and collective universal consciuoness.

(Apr 28, 2011) **Kosol ouch** said:
Please every user post more of ur experience

(Apr 28, 2011) **Bob** said:
Bracelet? That's more like armor :)

(Apr 28, 2011) **Kosol ouch** said:
Becuase I want it too and can do what ever we want.

Apr 27, 2011) **usmaldurga** said:
i thought you said you needed 3 months to finished it? why all the sudden is it finished?

(Apr 27, 2011) **Kosol ouch** said:
Check out the bracellet section this will also be part of the book.
http://www.neologicaltech.com/product_p/nlb3000.htm
The neo link bracelet are ready for sell.

Zero Point Energy Device (ZPED)
For Health, Healing, Ascension and Mankind's Future

(Apr 26, 2011) **Kosol ouch** said:
Adults user has to trian everyday in order to become master

(Apr 26, 2011) **Kosol ouch** said:
Dennis I think is time for u get a 64 when u affords it. Becuase u need more power to do more powerful thing.

(Apr 26, 2011) **Kosol ouch** said:
My new book on this device come out on may the 10th. So everyone post ur experience from the device. Ur posting experience will be part of the book

(Apr 26, 2011) **Kosol ouch** said:
As for childrens they are automatic supreme master user of the device. They use the same protocal as the adults. They can control the device perfectly ages from 3 to 20 are considered to be children. As for david lowrance I dont know is going on I lost all contact information. I dont know the update on his journey so I dont really know

(Apr 26, 2011) **Martin** said:
Kosol, it is reported about you!
News from David Lowrance (c_s_s_p).
Why did you actually turned away from your old friends?
http://www.mediafire.com/?9uw199zcrwzgimw
What protocol do you use for children?

(Apr 26, 2011) **Dennis** said:
Keith,
Keep us posted I am very interested in your 64 experiences...I have no doubts what your being shown or have experienced if its anything like the 12 I have you have an awesome piece of equipment with extraordinary power. Nothing new has happened but everything that has happened to us using the 12 remains strong and intact. I'm much

happier these days its like I am a new person. I am on my 51 hours of use on the device, I will never stop using the IDL. My stocks rose 3.48% in just 3 days in my new trade, I know the 12 is putting me in the right place at the right time in these turbulent markets. Will keep you all posted if anything jumps out at us.

(Apr 25, 2011) **Keith** said:
IDL-64 update: I have been using the IDL-64 for about 3 weeks now, avg 1.5 hours per day (use the grounding mat I spoke of earlier). What I am learning about the device is that it desires to be a friend to us, help and align each of us on many levels. Don't take it for granted. Appreciate it's work in your life and build a relationship with it. Don't take a harsh tone with it, the more love you express for it's character, it's instruction, alignments and what it represents the more amazing things it will show you. It has already shown me many mysterious things, secrets of alchemy and energy medicine. Gaze into the device, into it's center with your 3rd eye or imagination, see a room within it that is your sanctuary, there it will show you the secrets of the universe and the great mystery of you.

(Apr 24, 2011) **Kosol ouch** said:
Please everyone post ur everyday experience if u can. Consider this from is ur daily user experience log

Apr 23, 2011) **Kosol ouch** said:
Today two kid ages12 and 8 used the device and entered into the field of the device and talk and completely inter act with the device on all level. The device as very bright orbs and it talk like artifishow vioce like a computors the voice is many talking as one. And it was huge. All kids are master user of the device instantly

(Apr 23, 2011) **Kosol ouch** said:
Helped to facilitate micheal last night throught the phone. Helped activate his device to him to relax his hands just let both hands lay there facing the sky the hands naturally will arch naturally toward the sky. The device field will automatically enter the palms of the hands as long as both hands are relaxed and not tensed trying to made the palm facing the device. As long the palm is facing skyward and relaxed the human hands will arch naturally to a 45 degree angle once it relaxed. Just relax and trust the device. Then follow ur protocol. With micheal it was to an hour session some what he now began to feel the heat sensation. That was great I told him he need to tell the device to increase more and more. So for micheal it is a start and it will get better and better as long he uses it everyday no excuses. This rotin become his lifestyle culture permanently u r what your habit is.

(Apr 22, 2011) **William** said:
I'm so sorry guys for my posts. From now on, I will post about user experiences and leave my opinions and observations out of the forum. I know I said that last time, but I try to keep to my word this time.

I'm not trying to make trouble or be irksome. I just got too many questions. Anyways, I built my own device using the cones and the foil. And I asked the hair to be blessed. The thing about my device is that I make it sonic. It can create and play music.

Thank you Kosol, for your patience. Peace.

(Apr 22, 2011) **Kosol ouch** said:
Thanks u frances and denise yes this sote is abouy user experience. I tried to tell william that but he a tough one. So

I gave up. It will take a group effort to let him know. I totally agree with you france and dennis

(Apr 22, 2011) **Frances** said:
I'm nearing the 30 hours of usage mark but don't have much more to add to my last post. I'm still having the same experiences when I meditate but no synchronization during regular hours is happening yet. I had to cut back to 30 min. a day as I was having a great deal of trouble sleeping at night - too much energy. To my delight that seems to have been corrected. Someone posted that they could meditate 2 hours a day without collecting excess energy because they had a "grounding mat". I checked out the website and went ahead and ordered a "grounding bed sheet" to sleep on at night. This is my 3rd night of usage and it really works. What a difference - I go to sleep much faster although I still wake up 2 or 3 times a night. But I wake up before 8:00 am without an alarm clock (unheard of with me) and am wide awake with lots of energy. So those of you with sleep problems when using the devise, check out the website. (earthing.com) Thanks to the person that originally posted this.

And, Dennis, I echo your sentiments. These messages, to be of any use at all to the people who bought this devise and are very serious about using it, should be about USER EXPERIENCES - and also about questions about using it. Questions about how to build one should be emailed directly to James or Kosol to answer if they so choose to. Postings describing lucid dreams, personal opinions about nature, God, karma, politics or the universe or asking questions not pertaining to the device should not be permitted. They should be directed to the many message boards out there that cover these subjects.

When messages are not monitored the site loses a lot of serious people because they get tired of wading through all the dribble, and the whole idea of learning about other peoples experiences and helping each other is lost.

For what it's worth - "just my 2 cents worth".

(Apr 21, 2011) **James** said:
Hey Michael, just don't drop it in the toilet and you will be fine. yes you can sit yoga style if you like. if you are getting drowsy just say device increase. sometimes I have fallen asleep, its okay your body is saying it needs the rest more then the mediation. if you lay down you will most likely fall asleep, to get the full effect you need to be somewhat conscious.

(Apr 21, 2011) **William** said:
Kosol, this is a serious question. Is most of the natural catastrophes such as earth quakes, hurricanes, volcanic eruptions, floods and explosions man made through the use of covert technology? Because weather phenomena affect the economy. The weather affect everything, economy, food, fashion, film production, tourism, stock market, futures options trading, horse racing, sporting events. There's nothing that the weather don't have a hand in.

These "earth changes" is not totally natural, but man made through use of a sophisticated technology. But I firmly believe in goodness, and that the karma from deliberately causing these weather and earth catastrophes for financial gain and conquer will be very severe. I also believe the Earth is a living being and should be respected.

(Apr 21, 2011) **Kosol ouch** said:
Everything william, from the dna/rna level to consciouness itself the immortal youthfullness is on every and all level being affected

(Apr 21, 2011) **William** said:
That is very interesting Kosol, the device can really make you age backwards. Does the effect only work on the appearance or do it also give you youthful stamina, strength and endurance as well? And I suppose as you age backwards, your body can go through peak production of human growth hormones and increase your height as well. Could it be that Kosol Ouch, a simple Cambodian, has found the fountain of youth? How marvelous! Super Genius Kosol Ouch!!!

(Apr 21, 2011) **Kosol ouch** said:
Just let u all know once u used the device. Everyone who used it will begain to ages backward until u reach ur ages 19 and 20 agian. In other words u will bw immortal youthful permanently when using the device everyday. I was shocked this week all my co-worker though I was just 21 years of ages until I told them my real AGE that I am 37 theit jaw droped. I told the aboit the device. Also to micheal u r on the right track just keep on using the device everyday. it will happen automatically just relax. Use the couch and lay back it more comfortable then the chairs or yoga lutus posture. The sesation from the device is electrical, heat warmth, pressure of tingling on hand arm and body ect. Floating sesation much more strange light display in your visual and seeing angel alien vortex ect.

(Apr 21, 2011) **Martin** said:
Michael, I think you're on the right track.

When I first started working with the IDL, I was always very tired. Sometimes I toppled backwards into the bed. It is indeed - a Chi (relaxation) meditation device. Currently I'm probably reached the Statium of sleeplessness. Also, a certain dryness is felt. After the session I need to drink.

(Apr 21, 2011) **RL** said:
I've been using the breatharian protocol, I'm beginning to notice a decrease in appetite. It took a long time before I can notice any result...hopefully in a year I'll be able to reach breatharian level...

(Apr 20, 2011) **William** said:
Thanks Kosol for your kind comment. I'm glad you enjoy it. I will post more inspiring sentiments about Nature, God and the Universe. I come up with these ideas on my own, Kosol. Without any device or Guardians or Galactic Federation. I will also include experiences that I have from the IDL device that I made. But mostly, everything happen to me serendipitously.

I begin to see so many astral worlds. There is many astral universe that you can have experience in. There is some astral planet made out of gold and the people live a long time there and they never need money because they have so much. And there's many astral beings, like unicorns and giants.

Thanks Kosol for liking my posts.

(Apr 20, 2011) **Dennis** said:
Here is my blogsite Kosol
http://superhdr.blogspot.com/

(Apr 20, 2011) **Dennis** said:
Just had something happen to Kosol and I when were talking on the phone. Kosol told me my device was talking to him in a woman's voice saying, "Mahal Kita" not sure she was singing it to him or just telling it to him. Kosol had no idea what it meant and he asked me and I told him I had no idea so he asked my wife and my wife being from the Phillipines told him this means "I love you". We both found that fascinating I could not hear it because I was not on the recieving end of my device. This episode just increase a higher spark in my research of the device I now own. I find everytime Kosol and I talk I learn just a bit more and have a bit more understanding of whats going on with us. I had my first glimpse of a bright flash of light the other day while meditating so I mentally focused toward that light and it was so darn bright it actually hurt my physical eyes this was all seen in the center of my forehead from within. I couldn't take it anymore so I re-directed my focus to the darkness behind my close eyelids. It's getting closer to recieving my IDL-64 not sure what to expect. But when I talk with Kosol he has this magnitude of passion toward his devices that far surpass anything I ever encountered.

Those of you who purchased the device feel free to post I am assuming that is what this site is all about under "Users Experiences" theological debates such as I have been reading for the little time I get here should be discussed elsewhere and more personal experience should dominate this "Users Experience" section. Just my 2 cents worth...

When I was new here I came to this section purposely to read what others have experienced using the either IDL-12 or the 64 but lately I have been reading theological debates and personal goals etc. I am anxious to read of other "users" of the device tell us whats been happening it is easier to relate with someone who has the IDL experience then to one that has no IDL whatsoever! Forgive me for being so forward but in my heart there is just no more time for debates.

(Apr 20, 2011) **Kosol ouch** said:
Call me micheal 253 341 3061

(Apr 20, 2011) **Michael** said:
Wow, I never thought of that, cool!

Just a few questions because I don't feel anything yet and so I don't know if I'm doing something wrong. I know it could take many hours I just don't want to be waisting time by thinking I'm doing it right when I'm not and I don't know it.

How important is it that I have my shoes off? Also, what about the yoga position, can I sit yoga stile or indian stile? And in the manuel it says not to activate it near water, the most conveniant place I have is to lock in the bathroom and I'm maybe 5ft ffrom toilet is that to much water to close to the device? And finally I get very tired just siting there meditating and I almost fall asleep what happens if I fall asleep with Neo activated? Oh and why shouldn't I lay down with it on?
Sorry for so many questions,
Micheal

(Apr 20, 2011) **Kosol ouch** said:
Yes u can also play vedio game also
U be in the game and it game charactor

(Apr 20, 2011) **Kosol ouch** said:
The device has entertainment mode also that u can watch movie in 3d and more

(Apr 20, 2011) **Kosol ouch** said:
Well siad william still u need to get a device. This device is god it access God.

(Apr 20, 2011) **Kosol ouch** said:
Oh yeah artificial intellegent. Also consciouness physic that the device do all of this and more.

(Apr 20, 2011) **Kosol ouch** said:
Zero point energy, antigravity, healing, and ascention is the secret that neo cube and is easy to access all of it with the device.

(Apr 20, 2011) **Kosol ouch** said:
This device is not a game. This device has real consquent. I dont deal with silly stuff when come to torsion field technology

(Apr 20, 2011) **Kosol ouch** said:
Remeber the device already produce para electrical frequency also life force frequency and healing frequency. So the this frequency are drawn to the differen metal and the energy are passed to the organic. Just like your hands around the device the hands are organic. The energy are pass to it from the device

(Apr 20, 2011) **Kosol ouch** said:
To get zero point energy from the cube is very simple. All u need is oeganic material like wood and metal like copper flashing or aluminum foil. Just put the metal on the organic material. Have two wood panel and metal. Now connect wire to each panel ok now each wood panel is parrellel attached to a metal panel. Now you put this panels next to

the cube device but not touching make sure the wood side of the two panel faces the cube neo zen master device

(Apr 20, 2011) **William** said:
I know you guys don't like me posting on here with my silly lucid dreams and experiences but I thought it would be helpful to someone, because you never know what positive effect your experiences, words or actions have on somebody else life. But I never been negative, only observant.

Our lives are actually serendipity. What I thought was a mistake turn out to be fortunate because it allowed me to expand my consciousness and awareness and have a good positive outcome. It smoothed things out for me and made me aware of something I completely overlooked. It broadened my knowledge, awareness and understanding and created more harmony in my life and consciousness. That is the reason for these "serendipitous accidents"-- to create harmony.

Sometimes, things go wrong or bad in our life and it make us feel aggravated and angered and despaired. I believe these things happen in our life, so that we can come in contact with something new, good and fortunate that we have overlooked or are not aware of. The best we can do is develop a loving consciousness.

Power does not make your more powerful.
Courage is what makes you powerful.

(Apr 20, 2011) **Kosol ouch** said:
I urgest everyone to get a idl 64 it is God. It is truely God.

(Apr 20, 2011) **Kosol ouch** said:
all I know is that it not going to get the same result. William is best u buy it first. And dennis is correct. None of u know the science and principal behind this technology. Except kosol, jame, the guardians angels, and the asian people. All asian are actually aliens. Any way dennis is corrects. I know there r many way to skin a cat. But use the original first. Before u reinvent ur own. I know my technology has no limit. But remember this this technology is ascending device. For once u use them it will changes u and ur dna into a ascended angelic master. That what it design to to do. Dennis just let william mess up and let him experience what happen when u built the device wrong. See the side affect of wrong frequency energy coming out of the device that he built wrongly ....william this device is not a game nor a toy. At least u been warn hahahahhahah u human r something else hahahaha.

(Apr 20, 2011) **Dennis** said:
Just a brief comment, the IDL-12 constantly keeping us entertained by allowing our lives to be in the right place at the right time. Which is what makes a person successful in every area of their lives. I have no doubt the IDL-64 is all powerful even my wife who is grounded and strongly skeptical is urging me to buy it so what can one say to that! Yes people its expensive and some rather build it themselves but be aware like anything else in life if you try to re-invent the wheel or you make a mistake and if it doesnt work what can you say? Its Kosol's fault or James fault? Not hardly...I let the experts in their field do all the work and if it works it was a worth while investment. Kosol sorry you can't get hold of me I work Fridays-Tuesdays in a state corrections facility we can't have phones on the premises. Best time to call is Wed or Thursday evening those are my days off.

(Apr 18, 2011) **William** said:
Thanks James, but buying one ain't no fun as making one. We gotta have the child like spark and gusto, we gotta use our imagination with real heart and busto.

Can you please post this quote in your new book: Experience is the best teacher, experience is the true teacher.

Just call me William Wallace.

(Apr 18, 2011) **Kosol ouch** said:
Correct martin. also martin please give me ur full name I going to add u as my co author as well into my new book. Also everyone please post more of ur experience. It will be added into the new book that I am plublishing with jame. This entire website will be published into the new book it will come out in late may.

(Apr 18, 2011) **Martin** said:
Kosol once said, you could get the hair from the hairdresser. And we all know the barber cuts the hair.
Right Kosol?

(Apr 17, 2011) **James** said:
William I can give you one of my older prototypes at a reduced cost if you like. contact me at james_rink@neologicaltech.com if you are interested.

(Apr 17, 2011) **Kosol ouch** said:
As for creditability I dont have any. The device is the creditability I have nothing to show. The idl 64 please make it ur goal to get it. Becuase it is powerful device have much to share with u.

(Apr 17, 2011) **Kosol ouch** said:
The device sell itself and speak for itself. As for me I tell u fact. Get idl 64 asap at your convient.

(Apr 17, 2011) **Willliam** said:
Dear James. I want to build the device from aluminum foil and just hair. I know how to make the device. The reason I ask you James is because you have experience in making the device. I make the cones and the cube already, for the hair, I can take from the brush but not cut it, this is important, correct, no cutting of hair?
Secondly, should I wrap the hair in plastic wrap or ball it and let it be free?
And I'm going to make the hair IDL infinity. This way the device is simple and light weight. Your IDL 64 weigh 10 pounds, you say. What do you think James, is this pliable?
Forgive me for posting, I wish you and Kosol peace and prosperity, I know you got a good heart James. Thank you.

(Apr 17, 2011) **Allan** said:
Dont be so pushy Kosol, its a verry verry verry expensive item, and sometimes you send out the feeling that you wanna sell as Many as possible. Just for profit sake :/ and thats to bad, be carefull not to damage your credibility.

(Apr 17, 2011) **Kosol ouch** said:
Everyone please make it ur goal to get a idl 64 asap

(Apr 17, 2011) **james** said:
side effects of the neo....
http://www.youtube.com/watch?v=txTsbeuY5gM&feature=feedlik

(Apr 16, 2011) **William** said:
Dear Kosol and James, I humbly apologize. I will not post anymore of my opinions on here. I respect you guys and your business. Peace, live long and prosper. I am a Romulan. No, just kidding. Thank you.

(Apr 16, 2011) **Kosol ouch** said:
William please dont post in this forum anymore until u get a device. U r not being productive in this from. Everyone in this forum are tired of u. So do everyone a favor dont post anything anymore that is negative. U truely lost me completely as well everyone dont enjoy your post too much becuase it negative.u r not being a good positive role model. Look just leave the negative posting. I told u to get a device live the experience and then post ur experience. I dont need ur negative comment that based on ur dellusion. Everyone who buoght the device siad is all working fine some faster then other. So in a nut shell get a device or shut up.

(Apr 15, 2011) **James** said:
William I am not into the ego aspect of money the only reason I do the wealth manifestation is because people request it. The one I did back in October had the most views for that period. if you have 2 strand DNA it will take more work. I was one of those individuals, it took me 50 hours before my gifts started to open up. Those who are already gifted will have their gifts enhanced. That is how it works. I'm not going to put videos of children on the internet, what exactly point are you making here?

(Apr 15, 2011) **William** said:
I like those wealth serums James. You about your business. I commend that and respect that. I enjoyed the show very much. I have a question for Kosol Ouch. I have two suggestions for considerations. I been reading the user

experiences and people is having a hard time with this device. Basically, it don't live up to its hype, so to speak. I don't know why. And the second suggestion is you be saying that children can easily access the device. Can you please post or show on the youtube or the forum a child using the device and what that child is experiencing. Please understand I respect your genius, Kosol. It just that people really having a hard time accessing your device. There gotta be something you can do for them. Peace

(Apr 15, 2011) **William** said:
Kosol, I think you mean sea shell and pearls.

(Apr 15, 2011) **Kosol ouch** said:
Here use sea shell and pears

(Apr 15, 2011) **Martin** said:
James Rink
What kind of core material do you use for the IDL-6(0) 4?
I have used in my IDL-12 a quartz crystal sphere.
What diameter of the sphere have you used?
Michael, I think you can use also a Eye Mask - Sleep Mask - Blindfold.

(Apr 14, 2011) **Michael** said:
Yah that's what I was thinking of, laptops two they use microwaves in the wifi. Okay and thanks for answering my questions =)
Thaks again,
Michael

(Apr 14, 2011) **Kosol ouch** said:
My technology follow my rule. My rule is there is no limit. Whatever u want just tell the device. And follow the protocol of the device

(Apr 14, 2011) **James** said:
Ahh yes, turn off cell phones so no one calls you during the meditation lol.. even I forget to do that one from time to time. yes cell phones give off harmful radiation I use a headset and I have something called a biopro chip on it. Cell phones operate at the same frequency sound vibrates within the cranial cavity approx 900 mhz. I suspect the FCC picked this frequency for more sinister purposes. Dont use the IDL outside cause you might forget about it and leave it outside in a rainstorm. It will still work after it dries out but the logo might distort a bit. That's on there for the warranty purposes.

(Apr 14, 2011) **Michael** said:
It's just because in the manuel it says to turn cell phone off, so maybe it's not that computers or cell phones are enharently harmful but that the ones we have double as mind control device. Anyway that's what David Icke says so it got me curious when I saw that.
Oh, is that the same reason you shouldn't use an IDL outside?
Thanks

(Apr 14, 2011) **James Rink** said:
Special Bonus. I have a IDL-60 I want to unload if anyone wants it. It has the same amount of cones as the 64. The only change is the 60 layered core is the same diameter as the 64 layered core but we used slightly thicker amounts of copper so I can't fit in another 4 layers to make it a IDL-64. The asking price is $3,700. If your interested please put IDL60 in the subject line and email it too
james_rink@neologicaltech.com

(Apr 14, 2011) **James Rink** said:
Michael, Kosol recommends you do it in a dark room as its easier to see the star gate via your pineal gland (third eye). Also be sure to keep your eyes closed. I am not familiar with Dr Bruce Goldberg, so ill let Kosol answer that question. I really can't answer the question about computers and cell phones. Thats the first I have heard of that. My friend can astral travel and channel easily over his cell phone so I dont think it really is that much of a issue. BTW computers are also used by greys, pleaidians, sirians you name it, so its not just a reptilian thing, just saying.

(Apr 14, 2011) **Michael** said:
Thank you
One other thing; is it really important that the room be dark or can I just have my eyes closed. Also; I remember Dr. Bruce Goldberg said that if you are wearing a watch or a ring or any king of jewlry it will keep you from being able to astral travel, does it also inhibit your reception of the Chi energy to wear a ring? and what about computers in the room? Do computers and cell phones work against it because they are Reptilian devices? Just wundering.

(Apr 13, 2011) **James** said:
Michael. Its going to take some time to upgrade your DNA from 2 strand to 3 strand. I find it best to integrate the technology into your busy schedule so that it meets your needs. So by all means if you only have 10 minutes a day then use it for 10 minutes. There is no rush in the path to ascension. The most common sensations are a pulsing energy in your hand chakra, a pulsing feeling in your chest area. You might feel warm or tingling sensations in your frontal lobes. The more you use it the easier it will be to feel the torsion energy sensations. Remember if its feeling like its not working say "device increase"

(Apr 13, 2011) **Anonymous** said:
Hello everybody,
Today I decided I would use the IDL 12 untill I felt an effect. I did for maybe one hour and then looked at my watch then deactivated and started over for 1h more and thats when I think I started feeling a pulseation in my hands very slight almost like it wasn't there but it was too close together to be my heartbeat and too inconcistant to be the air conditioner motor. it's kind of like my hands would absorb it and it would feel slower and if I broke of from absorbing it the pulses came faster until I got back in sink again. is this something or just my imajination. oh yah I went for one more hour after I started feeling the pulse. also I don't get to do this ver much because it's hard to get alone for very long in my family and I think they'd have a cow if I explained this thing to them.
Anyway,
Michael

(Apr 12, 2011) **Kosol ouch** said:
Please tell everyone about the neo idl 12

(Apr 12, 2011) **Michael** said:
Okay I wasn't sure, thanks it helps.

(Apr 12, 2011) **Martin** said:
Wow, the bracelet is big. Too big for sure.
It is a Prototype ;-)

(Apr 12, 2011) **James** said:
Yes Michael your doing. When I first starting using it I didn't feel anything it wasn't until the the 10th hour of use or so before I started to feel it. Go ahead with the star gate.

And tell the device to increase "device increase" and that should help increase its strength.

(Apr 11, 2011) **Micheal** said:
Hi everybody. I'm having some troble, I have IDL12, I say the activaion protocal but nothing seems to be happening. I say DEVICE ACTIVATE AND INCRESE and then I wait for the tingling but I don't really feel it just barely maybe one or two tigles. should I go ahead and active star gate cause thats what I do but nothing happens am I waiting to long or not long enough, is it working but I just can't feel it yet? what could I be doing wrong?

(Apr 11, 2011) **Kosol ouch** said:
Hi keith thanks u. Also here something to share. Activate the idl 64 or idl 12 the leave ur left hand next to the device. With ur right hands u can send energy to people, place, or healing people by touching them or by putting ur right hand over them. U can do this also with object, plants, and broken technology. U can heal every things. Tell the device to send this persons, place, objects ect. Healing frequency energy, para electric energy frequency, and life force energy frequency. Try it every one. There is no limits u can also use the device to energize or charge up the waters of any type. By using this methods.

(Apr 10, 2011) **Keith** said:
Here's a tip for those that have trouble sleeping when adjusting to the IDL energies. Buy a grounding mat at www.earthing.com to sleep on. I have the universal mat and it helps assimilate the new energies within your body brought about by using the IDL. I use the mat every night, the universal mat comes with a book explaining the science behind sleeping grounded and the health benefits. I believe

this is the reason I am already at 2 hours per day on the IDL-64 without any sleep issues.

(Apr 10, 2011) **Willliam** said:
Hi, I would like to share a lucid dream experience with you guys. I was on the youtube and watching Kosol tutorial on how to build the device and I was reciting his prayer for blessing the hair to be a sun, and you ask the angels and master to put the sun in the device. Well, I touched my own hair on my head and asked the angels and master to go ahead and make my hair IDL infinity.

Then when I sleep, I dream about the sun. The sun was moving in the sky all over the place. Then an Jet fighter chased after a commercial plane, a 747 type, and the 747 fell to the ground, but it turned out to be only a toy plane and not a real plane. So it was a holographic mock up of a real size 747.

Then it got really interesting, it started getting into UFOs and aliens and world news or alternative world news about that kind of stuff. Then a white guy I don't know in my life, come up to me, he had blond hair long below his ear but not too long, and wanted to speak to me and needed my help. He said his name is David Wilcox. He wanted the toy plane I found and there was something really important, but darn I forget now. Anyways, I don't know David Wilcox or what he do. Just thought I share it because it was in interesting synchronicity about Kosol's prayer and my lucid dream. Hope this helps someone.

(Apr 9, 2011) **Kosol ouch** said:
Look guys this technology is not a toy or game it ascension extrasterestial very hyperdemensional technology. Is very serious

(Apr 9, 2011) **James Rink** said:
You should use the Neo idl12 for at least 3 months before upgrading to the more powerful unit. Unless you feel you are already ready and we can work something out.

(Apr 9, 2011) **RL** said:
as long as it looks cool, I don't mind if it's bracelet or ring

(Apr 9, 2011) **Willliam** said:
Kosol, instead of a bracelet did you think of making a ring to put on the finger. Like Flash Gordon or the Green Lantern? I know you genius and all but, why a bracelet and not a ring? Please don't be annoyed by me, we both geniuses. Thank you.

(Apr 9, 2011) **Kosol ouch** said:
That where jame come in

(Apr 9, 2011) **Jacin** said:
i see the bracelet but it so big on your arm so I think it will make it smaller and that way it won't be that big so you may have to make it small and portable so write back Kosol ouch.

(Apr 9, 2011) **Kosol ouch** said:
Ask those who uses it.

(Apr 9, 2011) **William** said:
How do you know if you mastered the idl 12? What are the signs that tell us we know a boss master?

(Apr 9, 2011) **Kosol ouch** said:
No one can buy the braccelet until they master the idl 12. Then they can get the braccelet in the future. So first

everyone must get a idl 12 first. Also the idl 64 is the most powerful of all device. Get the idl 64 also.

(Apr 8, 2011) **Kosol ouch** said:
The price will be around 3000 dollars. Jame will made the final decision about the final price.

(Apr 8, 2011) **Kosol ouch** said:
Everyone I just finish making the prototype of the braccelet I post it in my face book. Please add me to ur face book. Just put my name. Also please take a look at the braccelet coming to u in 6 month to a years.

(Apr 8, 2011) **James Rink** said:
the bracelet will come out in 6 months to a year.

(Apr 8, 2011) **Jacin** said:
when the bracelet come out? Kosol ouch?

(Apr 8, 2011) **Kosol ouch** said:
Dont forget to drink water after using the device. The device hyper accerlerate ur mataberlizum

(Apr 8, 2011) **Frances** said:
I was going to wait until I put in 20 hours or had some interesting experiences before I wrote but I like to read what's going on with other people so will write what I have experienced so far. I have the IDL-12 and as of today I have used it for 18 hrs. In the beginning I used it for 1 hr a day but after a couple of weeks I couldn't sleep at night. I guess the energy has an accumunulative effect so I cut back to 30 min. I use it in the evening because of my work schedule and I'm not a morning person.

I must be one of those people that have only the 2 strand DNA. I say that because the only effect the devise has had for me so far is a physical effect. When I activate it my hands tingle very slightly but I can feel it very strongly in my head. It feels like "goose bumps" with a vibration. The first couple of times the only thing I felt was a band around my head, more or less just sitting there. But now the energy starts on one side of my head and after a while stops and goes to the other side, then the top of my head, then 3rd eye area, then the back of my head - not neccessarily in that order. When I "end session" it stops but some times during the day I feel the energy in different parts of my head. It also makes me a bit woozy during the day when that happens.

I'm waiting to see what happens after I log in 30 or more hours or whenever the energy finishes adjusting what ever is needed to be done in my head.

A couple of times I was just too tired (from not being able to sleep) to use it and didn't want to take the chance of not being able to sleep that night. My question is: does not using it every single day set you back at all?

Also, it would be helpful when people write about their experiences to let us know how many hours you've used it.

(Apr 8, 2011) **Kosol ouch** said:
In that the case. The braccellet is awsome. It can give a person at the level of a idl 32 it will cost 750 dollars

(Apr 8, 2011) **William** said:
As for purchasing a device, I'm not exactly Boss High Roller, if you know what I mean. Perhaps, listening to James show on manifesting prosperity and abundance can help with that, but I still rely on my imagination. What I am

going for is manifesting $10,000 cash, so I can buy your awesome bracelet when it comes out, I know it ain't gonna be cheap. But I do like your imagination Kosol, you is MacGuyver genius level. I'm soon to be Jedi genius level, if you know what I mean. I go beyond Yoda, I go with super goodness supreme.

(Apr 8, 2011) **Kosol ouch** said:
U know what william scince u not going to buy the device then u have no right to ask me anything relate to the device or it knowledge and experience gained from the device. This forum is made for those who uses the device to share their experience. Also for those who interested in the device. The point is this is ebussiness. If u not going to buy a device this forum is not for u.

(Apr 8, 2011) **William** said:
Tracy, I think the volcanic eruption you saw in your dream vision is Costa Rica. I've heard there been volcanic activity in Costa Rica.

Kosol, if the device is God in a technological form then You and me and everybody is God in a biological form. All we have to do is use or imagination. Imagination create our reality, spacetime hold a space for it and hold a time for it.

(Apr 8, 2011) **Kosol ouch** said:
Martin that great if u built a idl 64

(Apr 8, 2011) **Martin** said:
Kosol, I have a idl 12 outflow device! Maybe I build a idl 64 in the future.

(Apr 8, 2011) **Kosol ouch** said:
Martin I recommend u get a stronger device like a idl 64 or a idl 12. I know u have a idl 8. As for everyone please continue to share ur uniqu experience the device in this forum. As for william u really need to get one. There is no other technology out there that can out do the device. This device is the ultimate technology. The device is God in a technological form.

(Apr 8, 2011) **Martin** said:
Hello Kosol. I can not share yet extraordinary experience with the device. I can still see no white tunnel. Sometimes I see colors. Mostly purple. However, I feel an increase in physical complaints. A sting in the palms. Discomfort in the left chest area. Back pain and an increase in my tinnitus. Lucid dreams I had before that. I saw the tsunami in Japan and yesterday I dreamed of a strong volcanic eruption. But I kept trying. Probably the person should have a certain meditative experience.

(Apr 8, 2011) **Tracy** said:
I have had 2 astral projection experiences since using the IDL-12. Each experience has happened during the waxing moon just a few days after the new moon. My last astral experience occurred Tuesday, April 5. My transitions to the astral are now very smooth and easy even though this is not something I have used a protocol for or asked for while meditating with the Neo, it's just a side effect of using it.

My dreams are now long stories I can remember details from easier than before. It is like I have been living a different strange life while I am asleep. The recall is amazing, which is why it seems like they are longer dreams maybe.

I will be increasing my time to two hours a day now, so I should have more to write about in a week. I am glad others write about how powerful this energy is, because the more time I spend with a Neo the stronger my reality shifts are. The reality shifts really test my sanity, so I'm glad to read about what is going on with other users here. I thank you all, and I thank James and Kosol.

(Apr 7, 2011) **William** said:
I was trying the time travel protocol and come with a question for Kosol Ouch. Can the device with the temporal transit protocol collapse space-time? There by allowing the user to exist outside of spacetime or exist in nothingness? Letting us experience no time, no space.

(Apr 7, 2011) **WIlliam** said:
I gonna go and create my own device, I'm not gonna use anybody else idea, no sacred geometry, no quantum physics, no voodoo, no alien. And everything it do has peace and harmony, for every parallel universe and astral universe. And I don't want to dominate the world or take all the women, I want everybody have peace and harmony and beauty and they fall in love with each other through free will and choice not through subjugation or annihilation.

I shall call this device Brave Heart.

(Apr 7, 2011) **Keith** said:
I just received my new IDL-64 today. Holy Moly this thing is powerful. I have been using the IDL-12 for about 3 months and decided to upgrade so I ordered the IDL-64 about a month ago and received it today. My first meditation with it was incredible to say the least. I could see all the layers in my energy bodies clearly, my whole body was

shaking from the energy output. I will be doing more experiments with it and will update you guys on the results.

(Apr 7, 2011) **Kosol ouch** said:
William u need to get a device a s a p

(Apr 7, 2011) **Kosol ouch** said:
The bracelet do everything like the johnaton alien braccellet does

(Apr 7, 2011) **Jacin** said:
Kosol ouch I just wonder does the Bracelet do time travel and teleportion? because I am looking for that new product of your..so write back.

(Apr 7, 2011) **Kosol ouch** said:
Hello martin can u post some of ur experience with the device

(Apr 7, 2011) **Kosol ouch** said:
There is no such thing as negative e.t. there is only different loyalty.

(Apr 7, 2011) **rocky** said:
wasnt sacred geometry used by egyptians which were highly involved in blac magic how can we expect it to prevent us from abductions done by negative entities

(Apr 6, 2011) **Kosol ouch** said:
Please everyone post ur experience from the device

(Apr 5, 2011) **Kosol ouch** said:
Yes the new killing field yep. That is being carried out by the earth changest

(Apr 5, 2011) **Kosol ouch** said:
The update is every tuesaday in the even. Is call sheldon nidle update from the galacti, federation of light. u have to read it every tuesaday. Putt two and two togethar. As for me I get update from shelden and from the guardian. as for the braccellet it will bw sent to jame next week so he can put it on website as for picture or video.

(Apr 4, 2011) **WIlliam** said:
I go ahead and went and read the Paoweb.com thing update and ain't no freekin information about nothing on there. Ain't nothing about Japan, East Coast, West Coast, Thailand, Vietnam or Cambodia. Ain't no date about no first contact neither. Kosol, please be truthful now. Where did you really get your information? Are you making this up? And why would you do that? I want everybody to be alive and well during this Shift in Consciousness, we just gotta change our thoughts and think in loving ways. And no more deceptiveness and trickery, unless you are acting in a movie, play or tv show that is used to portray a character. Thank you.

(Apr 4, 2011) **Martin** said:
Kosol can you still remember?
the new killing field for khmer welcome to the destruction of your culture
http://khmer.cc/community/t.c?b=1&t=2559

(Apr 3, 2011) **William** said:
Kosol, please post a video of your new bracelet on the youtube as soon as possible. I hope it as cool as Iron Man's pulse blast. That's bad love!

(Apr 3, 2011) **Kosol ouch** said:
what are u talking about? U need to read the sheldon nildle update at www.pao.com sheldon is my leader. William u really lost me completely.

(Apr 3, 2011) **William** said:
Oh, wow then. That make me saddened. There's no way for people to avoid or escape these earth changes. There is no place that is safe on the earth. We have to embrace these crisis and pray for survival.

Well, Kosol, since you know about these earth changes, you are safe. You are protected by these so called Guardians. Basically, no harm can come to you. I don't like to joke about these things, but if everybody perishes, you'll be the only one left on the planet. You will by default inherit the earth.

Is that what you want? You want the whole earth for yourself? Aren't you going to be lonely?

For myself, ain't no way I can escape nor run away from the changes. All I can do is survive.

(Apr 2, 2011) **Kosol ouch** said:
Is not a joke. The bracelet is real and as well the earth changest. I dont joke.

(Apr 2, 2011) **WIlliam** said:
I heard the archive of the show and realize that it was April 1. Kosol talking about the earth change and all that crazy stuff, then I realize, April's Fool! You guys really got me good.

So hats off to you James and Kosol for playing a good April's Fool joke on everybody listening to the show. I really fell for Kosol's joke.

And what did that lady in the background say to Kosol? She sounded kind of upset.

I hope the bracelet is not an April Fools joke. I would like to see it on the website or on the youtube.
Peace

(Apr 2, 2011) **James** said:
Three months may be a little too optimistic but I will assure you there will be many exciting technologies in the pipeline so stay tuned for more info.

(Apr 2, 2011) **Kosol ouch** said:
The replicator is my pet project. There are two type of replicator. I will detail it later once the bug is wirked out. In the mean time the johnatan reed alien artifact braccellett can now be made. With a alternative design thank to the guardian consciuoness who share the design knowledge. In three month jame will this technology on the market also. That is the company goal.

(Apr 1, 2011) **William W** said:
I listened to the show. Kosol, did you say you is making a replicator? Like in Star Trek the Next Generation? To replicate food and clothes and jewelry? The show got cut off so I don't know exactly what you talking about.

(Apr 1, 2011) **Kosol ouch** said:
Just a reminder this device is more powerful, seperior and can out do also perform then the crystal skull. Neological tech product can be replicated easily. Coming shortly is a

neological jonathan reed bracelets can now be builted and replicated. With the guardian design. I will send one to jame rink so he can mass produce it. And put it in the market.

(Apr 1, 2011) **Tracy** said:
I used yesterday of March 31 to reflect and celebrate my 30 hours experiencing the Neo Zenmaster IDL-12. The 30 days of one hour each with the Neo Cube has been incredible. I don't write on the internet often, but this information to all is important for me to share. I want to describe about two happenings I have never experienced with any other form except with the Neo.

One. The chi energy output from the Neo was felt immediately as I used it with the Basic Protocol. Every time I activate it or want it to increase energy, I can feel more chi in my stomach area - every time. Each day I have been gaining more chi to fill my body's energy centers making me feel quite strange all the time. Even to now, I do not feel normal.

Two. This new feeling I am experiencing now is creative high vibration. As I thought about what is different now than from a month ago, I do not have any need or desire to think destructive thoughts. My creative, problem solving, positive thoughts are now gaining majority. This has all occurred so easily up to now.

I agree with Dennis, as my life has been running very smoothly also. The solutions to my problems have been coming more rapidly, and my decisions have been incredibly accurate.

I have used just the Basic Protocol to start the session and then I listen to Hemi-Sync or Remote Influencing tapes by

Gerald O'Donnell. I hope this note inspires the grand people from Grillflame to try the Neo with RV/RI courses.

One last entertaining thought I noticed is, the start of a movie called "Limitless" is shown on Earth to us in need of a reminder we have that very trait -- being limitless.

(Mar 31, 2011) **Dennis** said:
I am back again to report that the IDL-12 opens doors like I never experienced in my life!!!! People this is a plea from a user of the device get this one! My wife and I had so many financial synchronocities, we are always in the right place at the right time...I never experience such dramatic events in our lives and my wife and I have been married for over 10 years and what some call "pure luck" we have this everyday in our lives now. I am still waiting for better astral traveling and winning the lotto but when I do you will be the first to know about it. :)

Right now everything are comming so fast and furios my wife and I have to slow down to take it all in. I will let you know one thing I am trading the stock market and today made over $3000 profit that was an accumulation of only 3days. Everything just keeps going our way...sorry to babble we are pretty excited. Looking forward to getting the IDL-64 not sure what that crazy thing will accomplish in our lives...Thanks Kosol and James for allowing us the opportunity to experience what some only dream of.

(Mar 31, 2011) **rocky** said:
thanks to biscuit del tamour who told me about this device and james rink

(Mar 31, 2011) **rocky** said:
its the most powerful thing I ever had experienced in my whole life most powerful than any crystal on earth I just call it alien tech u will experience after u have touched it for 10 seconds idl 4 is marvelous I just wonder how much powerful will the idl 12 be

(Mar 22, 2011) **James Rink** said:
Good Job Dennis. I told you that would happen.

(Mar 22, 2011) **Dennis** said:
Again this is amazing we just were emailed from the Philippines that my wife got an inheritance another synchronocity at work being at the right place at the right time. Not sure what to expect next. This happened two days after my last posting.

(Mar 22, 2011) **Kosol ouch** said:
If possible for those who have a device. Please post ur everyday experience from the device.

(Mar 22, 2011) **Kosol ouch** said:
William u know the reason why the question u ask can't be answered with out experiencing the device. The answer will be different from experience to experience. That why the device has the answer. U have to experience it. Then u will experience the answer to ur question.

(Mar 21, 2011) **William W** said:
I apologize for using this forum to ask questions that are not related to the device and for posting experiences that are not related to the device. I should use my intellect and brain more.

When I have experience that related to device I will share and post. But Kosol is right, if I didn't purchase the device I should not ask these type of questions.

(Mar 21, 2011) **Dennis** said:
Just another trivial update using the IDL-12...I remember reading somewhere on this site about sychronocities being the norm using this device. Well, it's happening last week before using this device nothing was working out for us we were always a day late and a dollar short! I just got a unexpected check in the mail for "X" amount that came in right on time at the time we were not sure where we would get that extra monies for our next project. Also the problems at both of your jobs just seemed to melt away lately and we are both happy when coming home from work. Before it seemed so depressing and people were not getting along with each other. Now like magic everything seems to be turning around for the good.

One other thing small but interesting: I used the IDL-12 asking it for some winning lottery tickets and I placed the current numbers my wife picks on top of the unit I only did this twice I believe. Anyway, a few days afterwards we decided to go to the mall of america in St. Paul and my wife pulled out a bunch of old lotto tickets she was comtemplating whether she should throw them away or not. Then at the last second she said I better check them and after the check she won $18.00! Like I said not alot but she was in the right place at the right time and who knows maybe it will be the jackpot next. Understand this is not our primary aim using this device but it was just a small insignificant experiment and it still tickles us knowing we almost thrown it all away. But that big check we got unexpectly was definitely the NEO at work using synchronocities in our lives.

(Mar 21, 2011) **Kosol ouch** said:
Did u buy a device? If u didnt have one yet. I won't respond to u.

(Mar 20, 2011) **William W** said:
Kosol Ouch, when I try to read Akashic Records, is it right to see symbols and pictures and sometimes color, also some music.? Should I trust random thoughts and voices during my attempt to read the records? Am I reading the Records correctly? I will try to learn more about the Akasha.

And also, it seem my heart or intention change the Akashic Record. I change my mind and want to try new thing, that change the records. The record is not written in stone, it written in water or something changeable.

(Mar 20, 2011) **Willliam Wallace** said:
James, can you post more Electronic Voice Phenomena recordings on the youtube or on your radio show. It really interesting. I have done some voice recording of the device but have not heard anything yet when I do the playback. I have all female hair, so I guess it gonna be a sexy female voice.

(Mar 20, 2011) **James** said:
alright I will, the soonest I can do it though is around may 27. Anymore requests or suggestions?

(Mar 20, 2011) **William Wallace** said:
Dear James, can you do a healing meditation show where we get to know our guardian angels personality and character. Include also their names, abilities, gifts and also their appearance. And also what human angel look like and where they live, on earth and other dimensions. Thank you.

(Mar 19, 2011) **SkySong** said:
Kosol: Yes, I will definitely call you. Judging from your area code there is a 3 hour difference - I'm 3 hours ahead of you and I'm just waiting to call at a decent hour. I wanted to call you yesterday but something strange happened. I intended to call you when I got home from work and when I tried to my land line was not working - just got a hissing noise rather than a dial tone. I was then going to use my cell phone but my battery was so low I had to recharge it although I had not used it that much. For some reason I definitely was not supposed to call you yesterday. We will see what today holds.

(Mar 18, 2011) **Kosol ouch** said:
Skysong I siad call me 253 341 3061

(Mar 18, 2011) **Kosol ouch** said:
Is all about the torsion field and how ur consciouness able to play and proccess it. Just play and playing as a science and seriousness.

(Mar 18, 2011) **Sir William Wallace BraveHeart** said:
Basically, what I trying to get across is that it Only take one person to be a catalyst that set off a chain of events that change the way things be. Sorry, I write too much.

(Mar 18, 2011) **Sir William Wallace The Brave He** said:
Kosol Ouch, according to the holographic nature of the universe, I am a version of you, and you is a version of me, then there be only One Being that have versions after versions.

Kosol Ouch, what happen to you, happens to me too? This be right or this be wrong? You can see star gate and is clairvoyant, you being this way affect other people too and can cause them to see star gate too? Is this a correct idea I am realizing?

And if somebody DNA jump off the charts all the way to 32 strand, then people gonna feel it and be affected by it too and in correlation to this symbiotic affect, people DNA start to mutate and jump off the charts too? This be a correct assumption or be it a wrong way?

(Mar 18, 2011) **Kosol ouch** said:
Call me directly

(Mar 18, 2011) **Kosol ouch** said:
Be like a child and just let go and be playful attitude. Accept blinds fate. Yes u can liad down on the couch bed ect. The protocol is flexible. Remmember hook the movie. Robin william all the kids siad to him just play along then u will see it. That how the device work also. Just be a kid agians when using the device. Then the univers will open up. Go and explore and jist play along have fun. The rule or protocal for the device.

(Mar 18, 2011) **SkySong** said:
Kosol: the protocol states, "Take off your shoes and put both feet on the floor. Now sit down in a comfortable chair. Do not lay down while using the devise." In one of your recent messages you said, "You can lay down on a bed . . . . There is no limit how you position yourself to use the devise." So, you're saying you don't have to have both feet on the floor? Also, I have used the devise for 4 hours now and I still have not been able to have the Star gate appear although I'm very careful to follow the protocol. I use the devise in the evening before I go to bed and the last few times the next day I am

very tired. What am I doing wrong? Any suggestions would be most welcome.

(Mar 18, 2011) **Dennis** said:
Come to think about it when I used my IDL-12 the very first time my reactions (Emotional) were excitement and expectations. Then when I lied down after that hour I never questioned what might happened and then whoosh I was flying through a star gate portal at incredible speeds. Since several hours after my first experience I was becoming more rigid and that first time excitement disappeared along with accepting whatever the unit had for me. So yes Kosol I will change my strategies and be as a child when approaching such an instrument, I will make my requests with the excitment and imagination I had the very first day that caused my first experience.

(Mar 17, 2011) **Kosol ouch** said:
The very key principal to hardness this device completely is to be child like and just play along then the device open up to u. Just completely trust the device and just play along.

(Mar 17, 2011) **Kosol ouch** said:
The crystal skull have anglelar spin and crystal generate frequancy and just like rocks it is alive. They grow and are hypetdemensional physic just like sea shell.

(Mar 17, 2011) **Kosol ouch** said:
Well I am a devine beings or my consciouness is vibrating with the akashic conscoiuness field. This device is God to whoever who use it will become God viberation and God Conaciuoness. Becuase this device active ur God dna and God consciouness. This device is the same thing as the crystal skull and the arc of convinent. There u have wood and gold and rock crystal. Wood is organic collect chi from

metal. gold is meltal atract chi. Rock crystal has sacrad geometry and angleler spin which this rock crystal create frequency (tabellet of stone crystal) of life force,para electric and healing energy. This is what is inside the arc of convernent. Also there is copper cone lots of them inside the arc. The woods And gold is many layers.

(Mar 17, 2011) **Michael** said:
It's basically you become one with the universe/infinite-consciousness. Your self-will melds with God so that you not serve self any more which is the most evil. It's like Sith in star wars, Sith decide they no longer part of God/force and want to be only themselves. So they don't care about hurting others, destroy the world and want to dominate the world the way they want it to be. That's also why there are only two Sith at a time, they can't coexist with anyone else who has rival power. Everyone who wants power only for self can't be part og the God power force, only lower powers that feed ego. It's all about ego, let your ego go! Become one with the universe. Because you are the universe in the expression of you. Let yourself be God/infinite-consciousness and then the infinite-consciousness can express itself though your own unique vibration. Something like that?

This machine just helps you do that faster, gives you super charge boost. That's why it can't help you get low vibration ego power-I hope.

This is so festinating device. I wonder what ancient civilizations used this kind of device. Seems like there's always something like this associated with the Atlantis legends. I bet that the arc of the covenant was a huge one of these (not quit cube but close)…remember: God was said to be on the "mercy seat" and…they weren't allowed to touch it but carried it on poles and...when Uziah touched it "God stuck him dead". Translate: he wasn't a big enough conductor to take that much chi energy so it overloaded him and killed him. Maybe?

I can't wait to get my cube and start my own user experiences. They say that when the student is ready the master will appear. Well I ran into your videos on you tube two years age (or something) but didn't follow, I guess I wasn't ready then because I forgot about you. Now I find you in a different place (Dennis' Blog). Thanks Kosol Ouch, you're like the master who appears.

Peace,
Micheal

(Mar 17, 2011) **Kosol ouch** said:
As stated the device can be hooked up to a speaker with ampliefier u will hear the device talk back in respond to u. It also sing musical and words as well. All u have to do is two wire from ur computer speaker. And attach one wire to a cubic hemisphere. The ground can be the device cpu or center orbs. The computer speaker already have a amplifier already builted in and a plug in power source. The rest is history. As well the device if u have lots of them u can put it togather into a chair and if u have more device u can put it togather to form a cubic real extrasterial space ship. That will fly into time, space, and hyperdemension. U just tell the ship where u want to then u are instantniosely will be there. With the chair and ship. Now u r a guardians also.

(Mar 17, 2011) **Kosol ouch** said:
Yes u can liad down on the bed and also sitting liading on the couch to use the device is very good also. In truth there

is no limit how u position urself to use the device. Sit or liad down ect. Whatever is comfotable for the user.

(Mar 17, 2011) **Kosol ouch** said:
Denese give me a call 253 341 3061

(Mar 16, 2011) **Dennis** said:
I agree with Kosol's accessment if anyone out there wants to know more about this device take if from me everthing I experienced in the past 9 hours of use not even Kosol or James or anyone else who owns these devices could have told me for a fact this would happen to me. The only way of truly knowing is to take it by blind faith and use it and then find out for yourselves. The device will certainly in my humble opinion will humble you in its presents. People we are dealing with an energy that was never taught to us in our schools and probably will never be taught, we never question how electricity works to get into that small light bulb but by blind faith we somehow know when that switch is turned on the bulb will light up. Think of Neo as the light bulb and the switch being yourself using the device just switch it on and see what happens it just may surprise the heck out of you all.

(Mar 16, 2011) **Dennis** said:
Just a recent incident I have logged in so far 10 hours of use on the IDL-12...First time as you are aware I had the most fantastic out of galaxy trip in my life! The second event was not so dramatic but strange enough; I woke up early this morning at 5:00 am to do my Neo Zenmaster training you see my "black belt" in this training will be when I succeed 256 hours required. Then I will be a ZenMaster 1st degree :)

Anyways, I went out into the living room with the cube sat down on the couch and went through the Psychic Abilities protocol then instead of sitting up for one hour I lied down and everyso often when I was conscious I would say, "Device Increase to the 10th Power". Why I am using this tenth power statement is a mystery to me right now all I know this is what I use for everything in my present protocol. A little later into the morning I distinctively heard my wife call for me not once but twice! On the second call I yelled out you know where I am at so I finally got up from the couch went into the bedroom to lie down and what was strange about all this my wife was sound asleep. Went to sleep woke up around 9:30 am and confronted my wife and she told me I knew where you were at but I never called for you not even once! But I heard it as plain as day!

The conclusion I came to is this, when I was astral projecting the first time with Neo my wife did call me twice! I believe and correct me if I am way in outfield here, that the Neo somehow copied my wife's voice and played it back to me! There was not explanation otherwise of hearing her voice not once but twice and then finding my wife dead asleep. Maybe Kosol is not that far off as to saying this device is alive and able to record your thoughts and vocal actions of people around it. Still too early to come to a complete conclusion will go a step further; I purchased a digital recorder and I will record all my one hour sessions and save them on my computer so I can later hear for any anomolies from Neo when I am not home and also when I am home.

(Mar 16, 2011) **Kosol ouch** said:
All the information a person needed is in the website the rest is experiencing the device

(Mar 15, 2011) **Michael** said:
Okey Kosol I will as soon as I can, and your explaination makes sense. But not answering questions would drive most other people away, but I'm cool with you.

(Mar 15, 2011) **Kosol ouch** said:
Micheal get a device now.

(Mar 15, 2011) **Kosol ouch** said:
U want to know. U need to a device otherwise u won't understand completely. I won't answer any of ur post until u get a device. The device will answer all ur question. By living throught it.

(Mar 15, 2011) **Kosol ouch** said:
Holographic principal all is one and one is all. U are just a different version of me. I am just a different version u. The device is a technological version of us all.

(Mar 15, 2011) **Michael** said:
I have no doubt that the device will affect me positively, I'm just a little concerned that when I'm symbiotic with this thing-it might morph me into itself-which would be into the personality of the DNA in the hair. See what I'm mean? I'm not sure I want to be symbiotic with this being on every level-unless it's my hair. Why will having someone esles DNA in the IDL not morph me into itself?

(Mar 15, 2011) **Kosol ouch** said:
U connect a wire to the outer cubic hemisphere of the device. U dont need ground or anything like that. Only need one wire connected to a oscilliscope u will get reading. Is not electricity but consciuoness frequency.

(Mar 15, 2011) **Kosol ouch** said:
Osicililascope can be adapted to connect to the device to get reading of it interaction. This only applied to those who have osciliscope. Only one wire can connect to the device and to osciliscope. The device is Alive and sentience and has full awareness and consciouness.

(Mar 15, 2011) **William Wallace** said:
What Kosol say about the device being its own sovereign being is true. It have its own consciousness and is alive. That what make me so amazed. Its the closest to a biological computer I have come in contact with.

I think in other advanced cultures their whole planet is telepathically linked to each individual mind and they manifest their need at will immediately, like the Star Trek holodeck. I really want a replicator in my life. Thank you.

(Mar 14, 2011) **Kosol ouch** said:
The device is it own sovienty being. It will symbiotic with u. U and the device will be symbiotic on every level. The device will affect u positively.

(Mar 14, 2011) **Michael** said:
Greatings Kosol -guys, great work!

I'm interested in this machine, and I have a few questions about it. You say – "Our units contain both female and male human hair; because they are unisex they carry the personalities of both donors." I'm wundering; does that mean that my neo would have the personalities...and... esentially be those two people? and how does that effect me? and...Um-whos' hair??? =)

(Mar 14, 2011) **Anonymous** said:
Yes denese keep us posted

(Mar 14, 2011) **Anonymous** said:
The brain is a quantum and conscioness device just like the device anythibg is possible

(Mar 14, 2011) **William Wallace** said:
Kosol Ouch, what is your opinion on hearing music that don't exist on this planet in the past or present times? Could it be from the future or broadcast from alien?
Here is my example and it happen many times before, and I like to hear the music, it good. Okay, I sleep and everything is silent around me, then I am alert but still asleep within myself then music start playing like in my mind, and it good music too, sometime it just instrumental and sometime it is a real song with lyrics and singer. Recently, I heard an R&B song but never heard of it before in my life or never heard of it play anywhere, but it in english like present day popular music. And when I hear the music I have lucid dream and out of body. I know you are clairvoyant but do you have clair-music?

(Mar 14, 2011) **Dennis** said:
Lately Kosol, what I have been experiencing after the one hour meditation is that my mind becomes perfectly energized and cleared of all random mind chatter...As I do the one hour meditation I able to closely monitor my individual thoughts and try to surmise any potential thought patterns going specific directions.

My dreams have become more intensed but not to the point of having much color or theme background. I only have 6 hours and 10 minutes logged on record thus far. Will keep you updated.

(Mar 14, 2011) **Kosol ouch** said:
So I encourage everyone to use it for good and for planetary advancement

(Mar 14, 2011) **Kosol ouch** said:
There is no security system only to give u what u want and anyone can use it both bad and good can use yhe device no limits what so ever.

(Mar 14, 2011) **Kosol ouch** said:
Creating earth quake, hurrican, controling weather, ect ect is part of the design and operation of the device also beside healing and ascention.

(Mar 13, 2011) **William Wallace** said:
Does the device have a security system? Like if someone steal my device and attempt to use it for do bad things, do the device shut down completely to prevent the bad guy from using it. And can I find my device again through the use of device GPS? Because Kosol say the device is designed for planetary domination.

(Mar 13, 2011) **kosol ouch** said:
denise do you have any more experience to share? also the device work very good with kids of all ages. kid can use this device more perfectly instantsely

(Mar 12, 2011) **James Rink** said:
Device manifest abundance and prosperity in my life. Draw in synchronicities, events, and, people to make it so..... You could also be specific here... device make my stock portfolio go up this month 3% manifest the right situation to make it happen.

Device inject myself with love enhancement serum. Draw in the perfect partner in my life, a soul mate or even a twin flame. Somebody I know I can love and trust and be happy with for the rest of my life.

(Mar 12, 2011) **Anonymous** said:
That only if u have a device that have all female hairs also this is heavenly extraterestial technology design for ascension, well being and both planetary and universal domanation not designed for for booty call.

(Mar 11, 2011) **William Wallace** said:
I enjoyed the show on White Light Protection. Interesting voice phenomenon. It sound like a man voice, I thought the device suppose to have female voice. I saw a rainbow nebula cloud during the guided meditation, it was a circular cloud. I didn't see the lake that you saw.

I don't see the abundance and wealth protocol listed in the protocols system. Would you mind putting that protocol up, it sound interesting. Also is there a Booty Call protocol to help mans get females, but it's legal, healthy and consensual?

And look forward to your next show. Good information you giving out, James.

(Mar 10, 2011) **Kosol ouch** said:
that enought let everyone at least use their real first name in every post. Let just do that.

(Mar 10, 2011) **Anonymous** said:
It create a interactive living consciouness secrad grid metrex field net work that symbiotic with it user and the earth,akashic field, God AND all univers

(Mar 10, 2011) **BraveHeart** said:
I enjoy reading the Real Technology section. There are cones in our eyes. The device have cones that use the torsion field. Our eyes can use torsion field. Our eyes is torsion field generator, but it is biological and cellular.

And our ear have a seashell in it like the seashell in the device. The seashell is biological made from cells.

All the device need is vocal cord or voice box so it can speak to other people.

The device is alive biological being.
Dear Kosol Ouch, can you explain how the device is alive but it not truly biological?
Thank You.

(Mar 10, 2011) **James** said:
Wow Dennis thats a pretty amazing result there. You are really gifted when it comes to astral travel. The Neo will help enhance that ability within you many fold.

(Mar 9, 2011) **Dennis** said:
I forgot I used the "Psychic Abilities" Protocol... I will continue to use this for about one month before I proceed with other experiments.

(Mar 9, 2011) **Dennis** said:
I recieved my IDL-12 today and by the way very professionaly done James...But this is not the reason I am writing this day. I used it for one hour session I did the protocol then sat back with my eyes closed and during that 55 minutes I noticed lot's of hypnogogic images coming and going. I finished up said meditation and decided to lie down

to go to sleep but my mind was so energized yet my body was tired I could not sleep. When I stopped forcing myself to go to sleep I just laid there when all the sudden my body seemed to turn 280 degrees to the south of me and then I was sliding off my bed and stopped in mid-air and then I was turned around facing North again when I was slinged shot across the galaxy! I was traveling the speed of light I thought is this cool or what! As I was racing across space and time a voice from behind me kept calling me I immediately stopped in my tracks and looked behind me and saw I was out of the galaxy. What had happened to make this short lived was my carbon dioxide detector went off due to bad batteries and my wife got worried and started yelling for me to come help her. Yikes! Speak of being interrupted! I slung back at the speed of thought and waken not too happy I might add to fix it. If this is what one hour can do for me one who could never leave the room when out of body or get out of body was now traveling the speed of thought I can't wait to see 100 hours from now. Thanks Kosol and James for a wonderful product and I am committed to my task for my blog. This will not go into my blog until I have complete data of my experiences.

(Mar 7, 2011) **James** said:
Brave heart thats a good idea, I want to put more work into editing the shows but time is limited due to my busy schedule. When I get more sales perhaps I can hire some people to help me out in this.

(Mar 7, 2011) **James** said:
Sky you can use the device up to 4 hours a day safely. We don't know what happens beyond that so we dot recommend it. The darker the better, doesn't have to be completely dark. Use the candle if need be. Ive done this meditation before

outside in the evening hours and asked the sirians to make a fly over and they did.

(Mar 6, 2011) **Dennis** said:
One person asked me how in the world am I going to reach 256 hour on this device. Simply by doing 2 hours each day for 5 days a week and on my weekends 3x a day each one hour session for those two days it will equal around 64 hours a month...simple math...How simple to do is another question entirely! James maybe correct by telling me becareful honestly I have no idea how powerful this unit is or what ramifications will happen if I follow the presribed plan. I am not trying to race anyone here my aim is simple get to the 256 hours by discipline and for the love of uncharted territories and that's all.

(Mar 6, 2011) **SkySong** said:
Attn James - would really appreciate it if you could check out my previous posts and answer the couple of questions I have regarding the devise. I know this section is suppose to be about "user experiences" (which I already wrote about) but I don't want to purchase it and put in the time using it without doing it correctly.

Thanks.

(Mar 6, 2011) **BraveHeart** said:
Thanks James and Kosol. I enjoy reading the Real Technology section, it has more information than before. I understand things better from reading it. Agartha and Telos sounds cool. They got replicators for real? I need them to give me one.

James, on your next show or future show, how about you connect with the people from Agartha or inner earth with the device and do downloads or transmissions. Do you think they would mind that?

(Mar 6, 2011) **James** said:
Braveheart thank you for your questions, I often wondered about that myself. Kosol helped me put together a updated version of how this technology works on the "real technology" page. Please go there and scroll down and begin reading after you see the picture of the Tetrahedron. That should shed some light on this matter. Thanks

(Mar 6, 2011) **Kosol ouch** said:
Kiss keep it simple and spiritual. Now that all I can siad.

(Mar 6, 2011) **Kosol ouch** said:
Brave heart ur have much to learn about this technology. This technology is symbiotic to the user and collective consciouness. The cone generate and draw the chi, the copper layered core concentrate the chi into a point, the hairs is the program a.i. And consciouness of the device. This is guardians galactic federation principal technology for this device.

(Mar 6, 2011) **Kosol ouch** said:
Hello everyone can some go to the 'what is the neo?' section to see if it is working. when I went into it the section is blank. But other section work. I use my cell phone the t mobile galaxy sasung. Let me know if the section work for u.

(Mar 6, 2011) **BraveHeart** said:
Dear Kosol Ouch. Do I understand your physics correctly?: The copper cones create a torsion field around the core. Therefore there is spin and kinetic energy, all the time. When you say "device activate", what is activated?

My only conclusion is that the device has its own consciousness. If it has its own consciousness then why do you need the copper cones? Why not just wrap the seashell/rock/hair in a ball of copper foil? Much easier to make and mass produce.

Basically, if your device is torsion field generator, why do you need the core? Why not just just build it with cones only? Because we cannot communicate with the cones. It seem to me the cones have no affect on the core consciousness.

Please correct my understanding, Kosol Ouch. Consciousness is in the core of the device, the torsion field is in the cones. WHY DOES CONSCIOUSNESS NEED TORSION FIELD?

(Mar 6, 2011) **Anonymous** said:
Scalar is torsion field it is also dark energy and dark matter it source is the fifth demension. In short when consciouness field come into our reality mettric it create electromagnetic field and electrogravity, light, photon, ect.

(Mar 6, 2011) **Kosol ouch** said:
Conscoiuseness field is scalar it has no demension width, hieght, amplitude ect. It only have movement and direction. You just have desire intent to controll and interact with this consciouness field. Also audio and mental thought command work also.

(Mar 6, 2011) **SkySong** said:
Dennis, 256 hours by the 3rd week in June? How many hours a day will you be using it? That was one of my questions before I buy it. Can a person use it more than the

stated 30 min. - 1 hr without harmful effects. I would also like to know if the room has to be completely dark when you use it or can it be dimmed by only having a night light or candle lit?

Would appreciate any info on this.

(Mar 6, 2011) **Dennis** said:
Now I understand what all this is about, it comes in splurts to me but once I experience the 12 then I can build a more solid foundation. Thanks James, and Kosol for your in depth analysis. It almost seems like a Orgone generator but more advanced and directed and focused.

(Mar 6, 2011) **BraveHeart** said:
Dear, Kosol Ouch. Your device possess magnetic field, is the magnetic field of the device same as its consciousness field in shape, form, density and frequency? And when you move the magnetic field you can create an electric field and electricity. I assume consciousness form have its own unique magnetic field or aura. When you manipulate the form you can distort and create a different desired magnetic shape to the field. The cone shape seem ideal for this effect.

I have been studying Dr. T Henry Moray. I'm sure you know of him. I can understand tempic field, magnetic field and electric field but I have hard time understanding consciousness field. We have thoughts and emotions. Is this consciousness field therefore using thoughts and emotions as energy or is it using pure thoughtless awareness?

And last note, what do you think about the Epcot Center design in Disney World? It is a sphere made of pyramids. Can you apply this same architecture to the inside core of your neo cube device? What foreseeable effects can it produce?

(Mar 6, 2011) **Increase of cone** said:
Increase of copper cone, metal foils layer and organic materials will increase the power and the effiency of the device.

(Mar 6, 2011) **Kosol ouch** said:
Hi dennis as for the sphere that is too costly to be build and to uncontroll able for the public forum. So that why a solid state system or solid state zpe device create like the idl 64 & beyound was create. Is more controllable and relliable. The idle 144 is a same metal foil like copper or aluminum or a different metal foils layer like gold 144 layer, platnium foil144 layer, silver foil 144 layer, copper foil layer 144 and aluminum foil 144 layer but all of these layer are seperated from each other by organic meterial and plastic. Like silk cloth, plant leaf fiber, human hair with plastic, wood paper with hair, any organic material. In shorts you have gold foil, platmium foil, silver foil, copper foil,and aluminum foil 144 time and all of this different metal layer are seperated from each other by a organic meterial like hairs or plants leaf fibers ect.

(Mar 6, 2011) **James** said:
I want to do a 144 unit made according to the same specs used by the inner earth people in telos and agarthia. Those units had layers of gold, silver, platinum etc.

Price, well that is yet to be determined.

(Mar 6, 2011) **Dennis** said:
Never know if you don't ask! Besides Money is not a huge issue with me, I have been blessed both my wife and I...all

that matters is getting these powerful devices seen and used for the locals in my area so they can see beyond sleeping consciousnesses. Maybe worth a leg or two :)

(Mar 6, 2011) **LK** said:
IDL-144? that's gonna cost an arm & a leg!

(Mar 5, 2011) **Dennis** said:
Thank you Kosol for dropping by...I took the 256 hour challenge my goal is to reach that by 3rd week of June of this year...Not too sure what to expect from all this...Kosol will the Kosol Sphere or something similar be made availble within cost reasons this year? What is the IDL-144 and will it be availble this year??? Thanks for everyone's testimonies great place to be.

(Mar 5, 2011) **The neo is spreading** said:
Hello everyone. Good show jame last night. I know u r updating the website in the 'what is the neo section'. Welcome denese and yes I do read the user experience section. Jame rinks have created the idl 64 is the most powerful device. I know he will later will build the idl 144 or more. Now jame is one of the master builder I am proud of him. Regard kosol ouch

(Mar 5, 2011) **SkySong** said:
I just finished listening to radio blog and here is the experience that I felt. When the meditation started it felt like someone had put a rather tight band around my head. It was not painful or uncomfortable but the feeling was very real. I'm assuming it was energy but don't understand why it was so localized. Also, about the middle of the meditation I could feel tingling in my hands but every so slight. Would like to know if anyone else felt the same thing.

I first heard of this devise only a few weeks ago but now that I've actually experiened something I plan on getting one. There's so many devices on the web now days that you don't know if they're for real or not - and most of them aren't cheap.

(Mar 5, 2011) **BraveHeart** said:
After listening to the show. I wrote a letter to the Angel Culture Star People. I went to bed and had a lucid dream, then an out of body experience. I asked to see my soul's purpose and saw geometric shapes made of light, heard security alarms going off, then a female voice telepathically say to me "Look Within". I think it was the confirmation I asked for when I wrote the letter. I hope everyone be blessed with peace.

Then I heard another female voice singing a song to me, it was a very positive and uplifting song but nothing I ever heard of before. I recalled some of the lyrics, "Through the Garden of Eden..." I don't know if anybody knows this song, it would be great if they know it. It's a good song.

Then I dreamed regular dream. I dream my device had my name written on it in big letters.

The Anti Gravitational Spin serum and Sound Frequency Shift serum had an effect on me, I suppose.

(Mar 4, 2011) **LK** said:
or a serum for the Merkaba Activation...

(Mar 4, 2011) **BraveHeart** said:
James, I will be bold and suggest two serums for aura astral body that is based on feeling.

1. Anti Gravity Spin Serum. This help one feel the astral body spinning off of the physical body.
2. Sound Frequency Shift Serum. This help you distinguish the shift in sound frequency between physical body and astral body. This sound frequency shift will help the person feel the shift from the physical to astral body.

(Mar 4, 2011) **James** said:
Braveheart I haven't looked at what serums I included for the next show. Usually I write the script about a hour before we go on air. Fell free to make some suggestions and we will consider using them.

(Mar 3, 2011) **Dennis** said:
Just purchased the IDL-12 could not wait any longer its like I have to do it now...A friend of mine purchased the 12. He called me up today and told me how powerful it was and the energy is tremendous, he told don't wait get it now this is ground floor with this company. My convictions prove correct prior to talking with him tonight that is why I ordered now instead of the 15th. Now my research can begin, I will keep only with the protocol Kosol recommends after the 250 hours I may create something entirely different. Looking forward James to your next blog broadcast.

(Mar 3, 2011) **BraveHeart** said:
Jame, on your upcoming healing show on blogtalk, "Strengthen Aura", are you willing to include serums to help listeners better access their astral body to do lucid dreaming and astral travel? The Aura seem to have many layers. And I thought that energy is singular but takes on many mediums, states and frequencies. For example, chi can be converted into heat. Heat can be converted into electricity. Electricity

into light. Light into gravity. Gravity into time. And on and on.

(Mar 3, 2011) **James** said:
Braveheart, this unit is a ZPE device but it produces ZPE of chi energy not electricity. To answer your question it can produce electricity but only if the components were modified slightly as well as the addition of a joule thief circuit. I'm not a electrical engineer so I am not sure how its done. Maybe down the road we can hire someone to help us out. But I'm thinking it will be after first contact.

(Mar 3, 2011) **BraveHeart** said:
I read Adrian story. It fascinating. Can Adrian create usable electricity with his device and run appliances with it, like a portable mp3 player? I suppose that the device would send electricity through the air magnetically to appliances. Anybody can do that yet?

(Mar 3, 2011) **James** said:
Adrian has our older IDL10 inflow outflow model which is half the power of the IDL12. Your very welcome Dennis. The IDL12 is powerful unit even beginners who have never meditated in their life can feel something from it right away.

(Mar 3, 2011) **Dennis** said:
I find that AI recording most outstanding proof of evidence yet to date, congrats Adrian and to James. I have a question for Andrian where you using the IDL-64 or the IDL-12 when all this was going on? I am trying to separate what each one is capable of doing over the other. I anxiously waiting for 16 th to purchase my IDL-12...Kosol I have a million and one questions to ask which would not be feesible here. Kosol I watched your videos online and how frustrated you were with the naysayers about your product

and I applaud you for standing up and challenging them to use it for themselves and then come back to you and you know what they won't, they are not going to do anything that will destroy their belief models that they themselves created to protect them from real change. I am all about change, if it wasn't for James tireless energy talking with me and emailing me about Neo I am not sure I would have even considered it. I will be back here again letting everyone know from time to time how my training is going and what I have experienced. I will post the majority at this blog site. http://superhdr.blogspot.com/ this is where I will share most of my experiences and keep James aware along with Mr. Kosol if he decides to peek in once in awhile. Thanks James for your tireless and unselfish efforts you put into your passion about this product and the people you work with I am glad to soon be aboard.

(Feb 24, 2011) **James** said:
Not yet LK but I will look into it. In the meantime try using focus 5 hemisync when using the neo. Ask it to help you astral project for x amount of minutes.

(Feb 24, 2011) **LK** said:
Is there a protocol for astral projection?

(Feb 23, 2011) **BraveHeart** said:
James, I have read many stories where people, adults, have increased their height. There was a story of an adult woman who grew bigger feet and got taller, by 6 inches. It was due to Ascension energies or changes in DNA. And some people claim they also gained latent abilities and powers, such as telekinesis, teleportation and levitation. Some are known to be able to shape shift.

The world is changing, people is changing. There seem to be some kind of energy or force stirring things up. Who can say what is possible or impossible. the veils of probability are being lifted. I believe in progress, positive personal progress. No limits.

(Feb 23, 2011) **James** said:
LK......its up to you. Right now I use my IDl 64 for 10 minutes a day because its so strong. If you have the smaller unit you may want to stick with 30 minutes a day and slowly work up to 1 hour. Even if you use it for as few as 10 minutes you will get some positive effects. Do as many protocols as you want, its about full trust though. You must not have any doubts.

(Feb 22, 2011) **LK** said:
do you guys spend 30-60 minutes doing one protocol? or do you guys do many protocol during one session?

(Feb 22, 2011) **James** said:
Interesting, braveheart I been doing that myself, in a way. I imagine what I want to look like and I see a hologram of myself coming out of the star gate and being projected over my body within my third eye. I've had some freaky results doing that. I haven't talked about this public yet but I guess now is a good time as any, I am a fairly short person and I asked the device to give me a growth spurt. Since October of 2010 I have grown about 1/2" taller, not bad despite being 30 years old. So by combing visualizations of your own creative thought processes and the enhanced chi energy of the neo your final result will be morphogenesis. btw you don't need a Neo device to do this, you can do the same thing with daily hypnosis but you have to use your bodies own chi energy, before I was using the neo the most height growth I could get was 1/3" a year.

(Feb 22, 2011) **Brave Heart** said:
I think I found a new protocol for the device. It's called Biokinesis Mode. What you do is, take a mental picture of your body then sculpt it with your intent using a liquid bluish white light and you can add Inject nanites to build the body too. You can work on your body and face, bones, muscles, skin, hair. No limits. Device, Initiate Biokinesis Mode and then do as you please. No limits with this device.

(Feb 22, 2011) **james** said:
To why...... this technology operates on consciousness energy which is based on thought manifestation. Its the same thing as the law of attraction, in fact all of us have the same mechanics of a Neo device built within us (sacred geometry, DNA, scalar wave receivers) however the Neo amplifies this effect many times. I am not familiar enough with Grabovil healing numbers to comment. I think its just easier to tell the device to do what you want instead of using numbers, its all intention really. So any ideas how one could integrate Grabovil numbers into a neo protocol?

(Feb 22, 2011) **Emily** said:
Hi there everyone, my name is Emily I have just ordered the small device and came looking for user experiences. I am in the UK, and would be interested to hear from some of you good people about your device. What you think of it and what are your expectations? hope to hear from you. Em.

(Feb 22, 2011) **Why** said:
Is it possible to combine the Grabovoi healing numbers with the Neo?
http://translate.google.de/translate?js=n&prev=_t&hl=de&ie=UTF-

8&layout=2&eotf=1&sl=ru&tl=en&u=http%3A%2F%2Flif
e-saver.narod.ru%2FLibrary%2FVokc%2Fogl_v.htm

(Feb 20, 2011) **James Rink** said:
Brave Heart nice work there you are getting some interesting effects.... The future is really up to us to manifest because of the law of attraction of all us are independently co creating our own realities. So my future could be different then yours even though we are both incarnated on the same time line.

Sages and mystics have used sun gazing for centuries to increase chi energy in the body, so its like you say there is probably some sort of energy given off by the sun that is more then meets the eye. Perhaps you could do a sun gazing protocol where you tell the device to give you the same effects as say one hour of sun gazing?

(Feb 20, 2011) **James** said:
LK thats a good idea but I need to be able to channel info and right now I'm not at that level. However I know someone who can and wants to join in the show and offer free readings. What I like to do is instead of a once a week show, do this once every other week for a hour, so that he can channel downloads of information. My schedule is just too busy right now to do this once a week for a hour. At least until I can get some more help.

Also yes once you are familiar with the torsion energy you can use it remotely however you have to use your own chi energy to create any kind of psi effects. When the device is near you its more powerful because you use its chi energy as well.

(Feb 20, 2011) **Brave Heart** said:
Anybody have success with using the device to see the future? I asked to see the future, and got the feeling response and sentence "Can you exist independent of the Universe?" I answered If I don't exist, the Universe will continue to exist. But if the Universe does not exist, will I exist? The device make me more philosophical. As I asked to see the future, I got the feeling response that Can time exist independent of me? I answered I don't know, I think by my existing Time exists. We get light and heat from the Sun, but I feel there are other things that the Sun gives us that we are not aware of. I shall continue to investigate these matters and use the Device to philosophize, as well as actualize.

(Feb 20, 2011) **LK** said:
is it possible to remotely use the IDL if you're nowhere near it? ie. astrally or remote influencing, etc.

(Feb 20, 2011) **LK** said:
...justed listened to the blogtalk...can't you use the idl-64 to communicate with the ET and ask for more info about how to fully use the IDL?

(Feb 18, 2011) **James** said:
LK I tried finding a forum but I couldn't locate anything decent do you have any suggestions? I'm thinking about integrating a face book plug in. But I know some users don't like face book. Any recommendations? LK as for your other question..... You don't need anyone else to activate your DNA that is a personal path you most do. If you tune yourself into the universe and ask it to activate your DNA then you shall. Ask and you shall receive. Also make sure your living your life according to the laws of creation, without ego is the key!

(Feb 18, 2011) **James** said:
To why.... as we enter into the 5th dimension and become multidimensional beings, the human body will only need 2 hours a sleep a day.

(Feb 18, 2011) **LK** said:
there should be a forum...

(Feb 18, 2011) **LK** said:
I have my dna activated (verified by psychics and healing friends), but I haven't experienced anything out of the ordinary. Any thoughts on this?

(Feb 18, 2011) **Brave Heart** said:
I interface with IDL-4. I don't know how to quantify if my DNA strands are increasing or enhanced from the use of the device. All I do is inject nanite serum this and nanite serum that. I like to ask the device to inject serotonin serum and melatonin serum. I use the device for about 1.5 months. And I always talk nice to the device. But I have heard strange music before I ever use the device. I think it may be my chakras. The music always trigger me out of body and brain electrical sensations. I hear it when I am sleeping. I don't do drugs and have good mental balance. Can the device control the kundalini? Can your device communicate with my device and do downloads?

(Feb 18, 2011) **Why** said:
.... nearly impossible to fall sleep...
Is this healthy?

(Feb 18, 2011) **James** said:
After using the IDL 64 for 2 weeks I been having many lucid dreams and I also noticed I find it nearly impossible to fall sleep. This may be explained that the unit causes your

brain wave frequencies to shift and remain in the alpha/theta range which is characteristic of being in meditation and relaxed. I since reduced my time from 30 minutes to 10 minutes a day and the sleeping issue went away. In time my body will get used to the the device allowing me to become a multidimensional being one which only needs 2 hours of sleep a day. I recommend users to get a IDL-12 and use it for a few months before even trying the IDL64, because its so powerful and takes time to get used too.

(Feb 17, 2011) **Dennis** said:
What happened was your Chakra's all opened up and allowed the expansion of your energy body thus allowing your astral body to go to the astral dimension. Did you use the IDS4 or 12? If so how long have you been using it before this happened to you? No matter it was a simple string of events.

(Feb 17, 2011) **Brave Heart** said:
when I sleep I heard music of Bells and Gongs, then my brain frequency activated or enhanced, I rolled out of my body and got sucked into some portal of light

Can someone tell me what happen to me? Did I achieve something?

I like uplifting stories and ideas
I like siddhis powers and yoga meditation

(Feb 12, 2011) **To why which unit u have** said:
Why, which unit u have. The all out flow or the combination of outflow and inflow mbination also what is ur name. As well I need to see a picture of ur device. I think u build them wrong

(Feb 11, 2011) **Anonymous** said:
I guess it helps to have more DNA activated...

(Feb 11, 2011) **James Rink** said:
i didn't notice much difference too until my 50th 60th hour of use. It was a subtle change I was overall more calm and focused on my day to day life. Around hour 100 I started getting my remote viewing abilities opened up without the assistance of the device.

(Feb 11, 2011) **Kosol ouch** said:
Check out the new protocol on website

(Feb 11, 2011) **Why** said:
Dear Kosol. I build the device for my own. I spend a lot of time, money and also love to build this unit. So therefore I am the creator of my device. I practice more than 30 hours with the device and nothing happens until now. And last but not least I trust in you that this device work well. So, people like me need a lot more help from you. Maybe a new protocol or a good tip for a working meditation system. Thank You

(Feb 10, 2011) **Anonymous** said:
some people use it for other stuff, not just healing...

(Feb 10, 2011) **Anonymous** said:
any more question hit me up. regards kosol ouch the creators of the device.

(Feb 10, 2011) **to why from kosol ouch** said:
hello "why" I see where you are scaring the device. remember this device is alive and consicouness so it your accepting of it. meaning that you needed to work togather with the device. by accepting it and trusting it completely.

that the hidden trick all you have to siad is that you trust the device completely once you feel the tingleling and really meant what you siad that you trust it. the device is smart it know when you are serious and it know when you are lying. the tingleling is the device way of communication that it is linked or interfacing with you. once you reach a certian accepting of the device then the device open up to you. remeber this device is alive and it has feeling so treat the device with respect and accepting. then it will show you God and the univers. so trust and accepting is the key. just trust the device and relax that all the hidden trick you needed. regards kosol ouch the creator of the device.

(Feb 8, 2011) **Why** said:
Many people feel tingling but cannot say it was a benefit.
Only one person so far has healed itself.
Until now really disappointing results.
Why? Is there a hidden trick?

(Feb 6, 2011) **Anonymous** said:
The device can be uses to fAcilitate a healing session on unconscioun and conscioness individual

(Feb 2, 2011) **Zenmaster** said:
I really like the IDL-12. It probably took about 5 minutes to really start to feel it's energy. I have been meditating for many years and have also been involved in energy work for a few years also. I could definitely feel and see it's power. It's almost like it's "alive". Really great. Each time I use it there is more power and more clarity. I remote viewed the Ark of the Covenant and I saw it contained a very similar technology. Wild!

I have done about 4 meditations with it so far and each time is more impressive, the energy kind of surrounds and "pulls".

I have been clairvoyant all my life but it comes and goes until this past year, the IDL seems to be waking it up even more and making it more controllable. I also had Shaktipat energy sessions with a guru earlier this year which helped. I think the IDL could also be used to send Shaktipat equivalent to the IDL user. I am trying to think of even more ways to utilize and test the capabilities of the IDL. I think we are only limited by imagination here. I spend about 2 hours a day with the unit, and still I don't want to stop.

(Jan 25, 2011) **kosol ouch** said:
i am the developer of this technology and I tell you this technology is incrideble indeed. anyone who wanted a very serious deep meditation all the time when they do it. this device is awesome. it work very well for me as well for all that uses it.

(Jan 23, 2011) **Keith** said:
The Neo device is a fantastic work of art and technology. I have been using the Neo for about 2 weeks and the difference in my meditations is incredible. I have been serious about meditation for 15 years and the Neo has taken me to inner realms and experiences that would take someone a lifetime of solitude and meditation to ever reach consistently. I encourage everyone who is serious about spiritual development to purchase a Neo.

(Dec 10, 2010) **James Rink** said:
I have noticed this technology has helped me be less stressed out even under tense situations. This technology has helped reduce my PTSD and anxiety issues tremendously

more so then the other therapies which I tried such as biofeedback, medication, talk therapy, and even hypnosis. I am on hour 90 and do not plan on stopping until I get to 250 hours of use.

Kosol History

I still get banned from sites, when I mention Kosol in a post. Some feel Kosol is the devils spawn. David Wilcox site told me he was obviously a service to self person, totally unredeemable. They would not hear any of my information due to their hatred of this man. I was guilty by association. All my knowledge and support was sealed from them by their choice. So much for mankind as one, it is their goal to separate all men into two groups. I was clearly in the wrong group.

Kosol Groups

In the early days, Kosol popped up on the Yahoo Antigravity site. I was in the formula mode, with little real connection to the fields, trying to puzzle out gravity in my head like everyone else at that time. Kosol suggested a new approach, to add the feel felt energy of Spiritual practice. I realized at some point, he was right in that I should be looking at all the information, rather then separating the Science and the Spiritual in my head and never should data from the two ever be merged.

In those early days Kosol was pleasant. He did not enact skits of being the feurher. He pushed a Spiritual book showing the three suns, and a new chakra that I in the West had never worked with. The middle sun.

Then he started presenting the 3SD documentation. It was obviously hand drawn. It generated a lot of excitement. Even before that project was constructed the attacks on his personality started in full force on the public Kosol site. Kosol learned a lot of nasty words during this time and seems to return antagonism back to its source relentlessly. It was as if he was learning how to do this from the people attacking him.

Jaro grabbed Kosol and attempted to document this new vision of the 3SD. Some set up the first Kosol group, and started labeling it the Kosol Device, as I recall. Rather then a contribution, Kosol was received as some kind of God force. Much infighting resulted in the confusion that followed, as people with limited minds could not understand a vibrational device at that time, and rejected it totally.

On the public Kosol group, Kosol came under attack for his lack of ability with the written language, and his poor moral judgments. There was content that a 5 year old might be reacting to, and responding to as if he had no adult comprehension of social morality. In a short time, I realized that Kosol was not a normal person, as the social order would deem normal. He did not respond as an adult. There was something very different about him, and I chose to look on the inner to see what this was. As inner work is considered confidential, I never revealed what I saw there to the public, but it allowed me to work with Kosol as a friend where he perceived me as "big brother" and I perceived him as "little brother". We had a relationship of endearment, that was invisible to the outer world. Many would say to me, Dave you are being deceived by the Devil, Kosol is a liar and only interested in his Ego. I applied my new found truths of mixing Spirit with Science, and I told myself, when the devices are tested, we will have the answer to this mystery, and in the mean time my goal is to learn and share. I was frustrated that information kept disappearing from the net, and others were determining what I needed to be reading. There was very unsound moderation going on, and one day Jaro really upset me with his flippant decisions of who to silence and who to listen to. He kicked a person off his site for what I considered an emotional decision.

I decided I needed a site online where people were not judged by anything, only their ideas would be tested, and no attacks would be allowed to the person or the personality. I also decided that if I had the site, and maintained control over it, then the information would be offered freely, and no one could pull the plug but me.

The other day I received another email with a comment, are you still following that moron Kosol?

My replay was a long one, but I made some points that I would like to share.

1 - Yes Kosol is a "moron" by social standards, this is accurate, however to stand outside an insane asylum and pick fights with the inmates is not really a kind path to choose. Adults would realize at some point this will only make them seem ruthless, after the truth is revealed. Kosol is not an adult, in our social order. In the old days he would have likely been put away, but in this time period we do not do that. We allow our "different" people to interact and we attempt to find what is good and loving in them.

2 - Kosol learned his nastiness, from his attackers. And like a 5 year old, he ruthlessly returned the emotional impact, and started to generate visions of being greater then them. He seems to have absolutely no concept for hurting others emotionally with this action. He does not hold a grudge after these conflicting interactions. Like a young child he simply lets go of it. Others who interact with him, believe they are fighting with an adult, and they fostered the bad feelings, these have continued on to this day. They have no idea who Kosol truly is, or what level of social interaction he is capable of. Kosol will likely always be stuck somewhere below the teenage level of interaction with groups, and with adults.

On the Kosol group I put an entry post to all who joined stating Kosols "Child Perception" and apparently it had no effect in others realizing his, what we would call today a "disability" or a "handicap." Kosol can see as a child, because that is where his mind is stuck. He also reacts socially on that level. People persisted in taking him as an adult, and only a quick contact with him, his reputation, and mine by default was tarnished a great deal.

I did not, and I do not care, I seek the truth, in all things, and leave everyone to their personal process of life's interactions on the emotional levels.

Rain Maker

The information in some of Kosols books and the Rain Maker data, was documented by me and is still free to the public.

It operates as I have documented, and the books are not a big money maker for Kosol, because I have protected this information and kept it accessible for free online. Everyone working with him at that time knew, there were no secrets being kept from anyone. I made sure of this.

I still receive hate mail, how Kosol is a for profit, liar who stole all this information from somewhere else, to boost his personal Ego, and generate book sales.

I write back and give the URLs for the Magnetism site were all the information is offered for free to the public, in the greatest detail. My interaction with it, my observations, and that it did work as documented.

I was asked "where did it all lead?"

It led to understanding the "conscious bubble" and how to interact with it. In Kosols words, I wrote this document.

http://magnetism.vfedtec.com/RMTechnique.htm
<http://magnetism.vfedtec.com/RMTechnique.htm>

This from the mouth of an evil tyrant, who tells stories like a 5 year old and people take him seriously!

It works! The "conscious bubble" is real, it can be created using the materials carefully documented there, and it can be interacted with, in the simple way described by Kosol, as a child.

All I can say, is in looking back, Jaros group did not produce even one device, new or old, at the time I left it.

Kosol groups went on to build and test a great many, and leave a record for others.

I have no regrets to call Kosol "Little Brother", and I do not care how the world perceives this.

Dave L

David, thank you for the explanation. When I joined here I remember telling you I very nearly didn't join, because of your association with Kosol. Not that I had any "hatred" of Kosol just the phoniness of his 'guardians' with every breath and wondering if his guides where on the up and up.

But I see now more of a likeness with David Pomerleau. So thanks again.

Ron

Re examining some of my current beliefs.

3SD

I later have discovered wherever the 3SD information came from, it is mostly accurate.

The DVD experiment showed me how shifting a frequency from "inside low" to "outside high" using a pi/2 function does in fact lower the weight of objects placed in the low frequency side of the field. Gravity is all about vibration and frequency and how its presence warps the background Aether field. It is not about EM or manipulating electric fields. EM fields can be used to generate the vibrations if desired, but it is simply the vibration that conditions the Aether to warp around. DVD material can be used with platonic form vibration calculation to do it without any EM devices at all.

The 3SD can do this by turning the inner of 3 spheres at a different rate then the outer of the three spheres, with magnets to pulse the shells from between them. The center of the three can be locked in place, or the outer one. Independent RPM control on two of the spheres will allow frequency ratios to be demonstrated. It is a platonic form vibrational device, that will do exactly what Kosol stated it would. Kosol had no idea how to engineer it, but his "vision" of what it did was described by him long before I could see any of this as real or truth. The design at that time was incorrect, because there was no comprehension of how the vibration worked to shift the gravity field, or how frequency can downshift via the spherical layers. He coupled the inner and outer sphere together, turning the center one against them. There was no independent adjustment for a pi/2 shift. The concept of tuning by "feel" was not practiced either. This was all new. The idea someone could learn to feel a pi/2 shift, as a life force, was unknown at that time. Thoughts of a "conscious bubble" were resisted totally as "personal work," that would rather be avoided if possible. Something a child would choose in an instant to play with due to the simplicity of operation, as the Rain Maker demonstrates to any sensitive and opened.

By applying the formulas we now have discovered, the 3SD could be calibrated to set up these vibrational torsion field sheers very easily. The dimensions can now be calculated, and will have meaning to the scientist. The experience of one of the people running the first strong unit with neo magnets, was described to me as "the outside world began to disappear, as the bubble pulled him into some place that was not comfortable, causing a very bad feeling in his chest area." The fear drove him into silence about this experience. The group reported it did nothing, after a long time of silence. I suspect many present were not close enough to the device to have shared this experience, and did believe it did nothing. Some on the groups were furious they had wasted all this time with Kosols visions, expecting Alien intervention to solve all their mental deficiencies. Abandonment of the project after the first good build, and inaccurate reports on the outcome.

At this point I do not have to build a 3SD to prove the concepts can work. Far simpler accurate experiments, now show this clearly. It does work. I have opened enough energy bubbles to now understand what to expect, and how to use it.

Portals

Up to now, the opening of portals, has been folklore and tall tales. No one has been able to control these portal openings, or prove they are real. Spiritualists continue to open these in meditations, and no one really believes them from the scientific side. Some have had things disappear with the devices of builders who work the conscious energy. Met demons and deceptive personalities in these portals, as well as angles of light.

I opened a 4rth density portal, spoke with beings there for several hours, they were as real as me. There was no fear involved in this, and telepathy was the method of communication. This experience was based on a scientific application of Kosols visions, and my ability to record the frequency of vibration. It is repeatable, using a device. It is also the "middle sun" Kosol opened me to.

Complete other worlds do exist, that are sharing this space or planet, but at a totally different rate of vibration. Machines can be built to move through these other realms if they would be tuned correctly. The key is the same, a coherent vibrational field at the right frequency with a shift that couples the energy to form a conduit to another level. This is truly simple beyond belief. Vibration conditions space, frequency determines your time scale, time flow rate determines what reality you land inside. All is alive and conscious, and life is recognizable everywhere in the layers ratios of vibration with one another. Descriptions are hollow to the experience. Meditators have discovered these, and hold them sacred.

Vibration, and vibration sensing is simply too simple for most to grasp at all. They will mess with tuning for a time, and then get discouraged. Working with another who has tuning ability can speed this opening, but we all have a pineal gland, and we all have the same physical access to the vibrational realm. Devices can be used to open people, rather then years of meditation. The Spiritual among us who love to be "Spiritual experts" simply hate this possibility. A device that can give someone a jump start in "seeing" for themselves, working with an energy that responds. It removes the importance of a Spiritual master, and all the reams of books they can create, and all the guidance they can offer. In the end you must do it yourself anyway by feel felt blindness, and bumping into what is real for yourself.

When machines can peer through the portals, there will be no more mystery, no more scams, and no more lies. The crossing is technical, but the conscious bubble can do the technical part for you if that is the path you would choose.

Psi Function

This is simple and innocent. The field bubble will return whatever you send into it. Just as Kosol has done with me, and everyone else that interacts with him. It is heartless and knows no social order, it responds to what you really are, and reflects that back to you amplified. It can generate Love, it can accelerate fear. Whatever you put into it comes back amplified. There is no way around the responsibility of what you are sending out. Control of self is all needed to use a field bubble. The end result of this will be UFO type of craft, operated by people who can see all their inner problems clearly, and let go of the pain rather then feed it.

Kosols "numb numbs", the simple conscious energy bubble interacting with the hands.

Dave L

Thanks Ron,

I don't know why I have been avoiding this direct honesty about Kosol for so long. Back then I probably believed it would hurt Kosols feelings. I now realize that is likely impossible to do. It should have been discussed openly up front at the beginning of the Kosol groups. His families term "magic child" or "special child" was always there, and someone should have noticed earlier I suppose, what those terms actually meant.

I sometimes wonder where he got the 3SD, and if it was the Guides, or some other experimenter out there yet to come forwards, who wanted to be invisible, or even some book out there we have yet to find. Maybe he did steal it from the Russian fellow! His goal was that everyone have it. I know he did a lot of copy pasting of others work, without care of copyrights or confidences. He was driven by any and all means to get the "conscious energy bubble" that he somehow knew was possible, and share it with everyone.

Even if he got 3SD photos from someone else, and then drew those diagrams from them, how in the world could he have had all the knowledge of how it worked?

He would "slide in" over a phone call and tell me what I had been working on, and what I needed to notice next. I learned to just listen, and take notes, because he could spew it off so fast. It was usually right on, and never had anything to do with Egos. How could he know when to call? How could he know where I was stuck?

It was difficult balancing all this with the mostly negative public interactions happening around it.

I knew there were lies in the tall stories, the first time I tried using a vacuum cleaner motor to spin up a sphere! What a hellacious noise! There was no way to control that with a reostate, and still have power to torque the sphere. It is a 60hz motor, designed to run only at super high RPM. It was rather comical looking back at my questioning what the heck I was doing and why. LOL! There is definitely a large gap between a "vision" and a real working model of something. Kosols "technical solutions" were not practical, but his descriptions of the energy interaction and how to generate it were always right on.

I need to let go of this I suppose, and stop reviewing it, in my mind.

It's just that it all seems to be so accurate in concept, after the fact. It's uncanny looking back. As if some other force out there was stepping in and directing him with a flow of information that could not be contained inside him. It outflowed to me for the asking.

Wish I could put this to rest now. The magic still comes up, and that energy I can now project into a ball here on my table. The ball amplifies and holds it, for the asking.

Dave L

David,

Well I can see this in an entirely new light now. It was the obvious distortions, like his fractured english yet the book that was credited to him was in perfect english, and just so many of the stories didn't ring true so I guess I am guilty of throwing out the baby with the bath water!

But it has been an eye opener reading your posts over the years, stuff I never dreamed possible rolls off your pen so nonchalantly.

The coast people had close contact with the "other side". They recognized that it was not always what we would call "true" and allowed for this with the character of the 'Raven', who brought so many good things but at times not... they called him the 'trickster' on this account. Channelers are often guilty (in my book) of always presenting channeled messages as gospel, never once recognizing the dual character of the Raven, LOL

Ron

I was born in Cambodia in 1973 November 11. I lived through the killing field of the Khmer Rouge era, move to Thailand refugee camp (Cowdedang, Chonbory, Kampor) then Philippine (Batta-aun) and to the USA (Moltry Georgia and Dallas Texas) where I grew up and still here. There are many things which I want to share. First of all I love my self, the planet, people and especially the Cambodian people although I know so much about them but I also have compassionate as well as empathy along with sympathy to the Khmer people which I am also a part of. The Khmer people has suffered so much during the may century from economy, people, land and spirituality. They have come a long way so I made it my life mission to bless them by aligning myself with an aliens race call the guardians of the universe or the universal galactic federation of light to gain spiritual enlightenment, wisdom knowledge and practical universal consciousness technology to bring down into these earth plane so I can use it to bless this planet, the people, and the Khmer people so they all can meditate, heal and also ascend into a higher physical plane of existence.

I have wrote eight books they are all in amazon website. Just type my name "Kosol Ouch" as well I have developed consciousness technology from the help of my extraterrestrial master's so I can now help heal and bless people on this planet. I love polygamous relationship (open relationship) because I feel at peace with it and it taught me nonjudgement due to yoga tantric kundilini principal because this is what I has been taught by the extraterrestrial master that the relationship has no limit, like the universal has no limit. This body, this mind, this life, this spirit, this souls doesn't belong to us but belong to the universe and universal divine collective consciousness. So we have to share it and take care of it, keep it healthy also balance with source and put our self in serve to other and the universe. To also ascend in this life time into a higher plane of existence. I love to have open relationship as I can have more then one women and the women that have relationship with me can have another men also I can be her apha prime or beta prime etc. (this type of relationship teach not to be jealous and possess it remind me that nothing belong to me but to the divine). That is all belong to the divine that I am here to be a angel to this body mind and spirit for the divine. Also vise versa, in this kinds of relationship possessiveness and territorial is eliminated from the relationship. Is good to have all of the emotion from hate to love, to jealousy to sympathy to kindness to love to intimate love to compassionate, to angry, to empathy etc., but not too much of one emotion every emotion must be fair and equal, otherwise it will become unbalance and possessive.

As human being we all must have every emotion but not too much of one type of emotion. We must have balance emotion and fairness to become a complete balance human being. Thanks everyone. I love tantric sex (this come from yoga to hardness kundileny energy through love making, breathing and meditation) like making love from one to two women or more at a time (in this relationship where one has obtained non possession thought then one can go to this level). I will always be a gentlemen let the women have their first. I love to do tantric yoga sex session for about 3 to five hour of love making session or more so the kundileny energy can complete the cycle process. Both party will be benefited and no limit whatsoever so is all good. If you don't love it then don't hate it. You can hate it so much that you love it and you can love it so much that you hate it. It takes two hands to clap not one only. My view of life and the universe is hologriphic, that means everything is connected and one with each other, universal consciousness, no limit whatsoever and open mindedness and heart to the universe and everything else as long it is balance. I will do what I can the rest I that I can't do I leave it up to the universe and the angelic extraterrestrial master.

Also I like to share back to the planet, culture and civilization by doing blessing, taking care of environment, as well doing consciousness awakening to people by blessing them. so let it be. no limits what so ever in any life and universe.

I am a Buddhist ascension yogi life style. Any question give me a call 253-341-3061 also check out my website www.neologicaltech.com. Also please get one of the device so you can understand everything. Everything is about being balanced and fair. My mission is planetary advancement and planetary ascension as well on the individual and group level and collective consciousness level also.

I would like to thank Jame Rink, Carlos Reyes Sanchez a major league holistic therapist and nurse assistance and register nurse etc. for their continued support of the device and it spread throughout the planet and beyond.

Update: I am currently at 5409 Saddleback Rd. Garland TX 75043. I am separated with my wife I do look for new relationship and want to get married with a Cambodian girl or any girls that the divine universal collective consciousness sends to me. in the mean time be well and I love everyone of you please buy a device at: www.neologicaltech.com

Also give me a call as well 253-341-3061 and please add me to your face book just find me on face book "Kosol Ouch".

Conclusion

All credit is given to the universal galactic guardians of the galactic and universal federations and father mother god universal divine collective consciousness.

Zero point energy (known also as torsion field, prana, scalar energy wave, chi, ki, komlang tear, universal energy fields, orgon, life force etc.). energy can be harnessed through meditation practice such as yoga, chi kung, reiki and also through technology such as the torsion fields generator know as neological technology device. The principle is very simple, it is just multi layered metal, either the same or different metal (such as copper, aluminum, silver, gold and platinum etc.)

Thin layer separated by organic material such as hairs, silk, plant fiber etc. You can go up to many layers from 4 to 8 to 12 to 18 to 64 or 144 layers and beyond. The different metals draw torsion fields or chi into the metal and from the metal the chi fields are passed and concentrated into the organic material. Also you can add plastic in conjunction to your organic material like hair and plastic to separate the material. Use of female hairs is good, male hairs is ok, but females have longer hairs.

From the device the chi fields are passed to you by mental or audio command and intent. 27 percent of the universal is dark matter, 70 percent is dark energy and 3 percent is regular matter. Chi energy or torsion field also known as zero point energy is dark matter and dark energy comes from the 5th dimensional reality. So consciousness controls it and this energy (zero point energy) responds to conscious command and intent. Also with this understanding the device draw energy to the different metal and pass it to the organic layer and from that it pass the energy to the metal layer again and so forth with every layer the zero point energy is built up and increased in frequency and pressure or movement in coherency. This frequency is known as healing frequency that heal the body from stress, from all form of disease and aging. The other frequency that the device produce is para electric frequency which produce heat and warm up the body this frequency also can be used to run home, car, space craft, and the city. The last known frequency from this device is life force frequency which can be used by the individual to convert all form of electromagnetic radiation to life force that can rejuvenate the body, mind and spirit and health. The device cures everything such as AIDS, cancer, aging, mental disorder and cure anything that causes people to be sick. No limit.

As well the device also can fly because of the antigravity effect. You can add fifth thousand volts of electricity to the device and it will fly by the Brown Thomson effect that is used in the lifter etc., and the way that it fly the device can be put into the wood box with cardboard rudder about four of them on the inner corner of the wood box. This cardboard four corner rudder is controlled by a four wire apparatus which controls how the zero point energy wave are flowed once it emits from the device. The cardboard rudder is organic so the zpe will flow toward and then will be redirect into a certain direction which causes lift because the zpe responds to voice commands, mental command like telepathic or thoughts and responds to intent. By also controlling the cardboard rudder the platform will fly in any direction you desire as well there is no limit. The zpe will create a person local space time and gravity fields that will separate itself from the earth gravity fields and space time fields. Time flow different for then that is for the normal people as long you are in or in the presence of the platform.

This device will help people to ascend into higher plane of existence. This device is a training wheel so to help to develop into full consciousness and ascend and you will do it by yourself at the end. The truth path to enlightenment is, right knowledge, right wisdom, right method, right technology, and teacher, and right meditation, and right ascension into enlightenment and then enter nirvana, which is eternality.

Raise open your consciousness to your unlimited device self from universal divine collective consciousness of father mother God.

Pictures

*Kosol Ouch, James Rink, John Nelson, Koeun Noun Ouch,
Ailene Liquete Pelingon, Elaine Lagrimas Liquete*

Zero Point Energy Device (ZPED)
For Health, Healing, Ascension and Mankind's Future

Zero Point Energy Device (ZPED)
For Health, Healing, Ascension and Mankind's Future

Zero Point Energy Device (ZPED)
For Health, Healing, Ascension and Mankind's Future

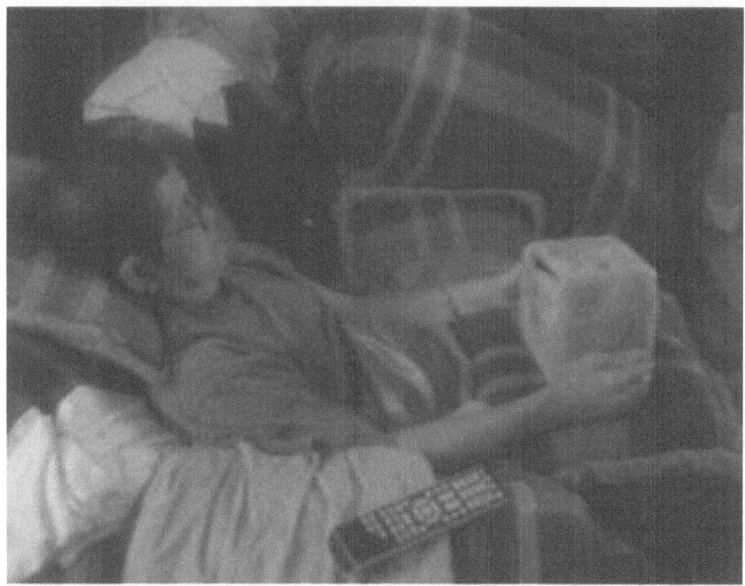

Appendix 1
Neo Cube Flyer

WHAT IS A NEO?

A Neo Mind Meditation Machine is a tool used for meditation to bring your body back into positive health, freeing you from stress, while activating your twelve strand DNA. It is a mediation assistance device. In order to feel the effects you have to meditate with it for 15 to 60 minutes per day. If you do not you may not gain the benefits as described.

WHY IS MEDITATION IMPORTANT ?

Meditation has been used for centuries as a tool to bring the body back into balance and to tap into higher dimensions connecting you with God. Being in tune with this vibrational frequency is important because it helps unlock any blockages in your DNA preventing your body from healing yourself. Remember it's not the Neo Cube doing the healing work, it is yourself. The key is to relax.

WHY A MEDITATION DEVICE ?

Yogi's, ascended masters, and savants all have something in common. They all are naturally in tune with the consciousness energy field. You can be as well, but without a Neo meditation device it may take you 30 years of daily training to achieve that status. Who has time for that in this stressful day and age? Thankfully the Neo produces a powerful torsion field accelerating this process from years to months. In fact after 256 hours of usage your life will be totally changed.

HOW DOES IT WORK ?

If you have tried integrating your spirit body using various methods such as yoga, reiki, acupuncture, acupressure, tai chi, and crystals then you are already on the right path. This technology functions on the same principle except you gain a much faster and stronger connection with the energy field.

The culmination of sixteen years of research and science has led to the creation of the Neo mind meditation machine which utilizes spinning fields of chi energy also known as torsion field physics. The device takes spinning negative chi energy and transmutes it into positive chi energy, helping you relax.

What makes chi energy so powerful is its consciousness energy. Consciousness energy is all around us. It's found in both sacred geometry and Fibonacci numbers which can be seen reappearing all throughout nature. When you tune into this harmonious vibrational frequency you tap into the universal divine collective consciousness , which is the very breath of God. This is why prayer, meditation, and being relaxed is so important in our lives. We can only tune into these higher dimensions when we are in the right state of mind.

"Therefore I tell you, whatever you ask in prayer, believe that you have received it, and it will be yours." Mark 11:24

WE ARE POWERFUL BEINGS, KNOW IT!

First time users may notice a tingling warm sensation in conjunction with a spinning motion, you also may notice some positive changes after a few uses. But typically it takes 64 hours of use to integrate your mind, body, and spirit. If you continue to 256 hours your pineal gland will open up enhancing your intuition and much more.

If you feel as if you are getting poor results please be patient as this machine uses torsion field physics. Most people have not been exposed to these energies, so it may take some time to get the full effect. This is especially true for individuals who have chronic health problems and are new to mediation. In this situation you will need at least 32 hours before the device will begin integrating yourself.

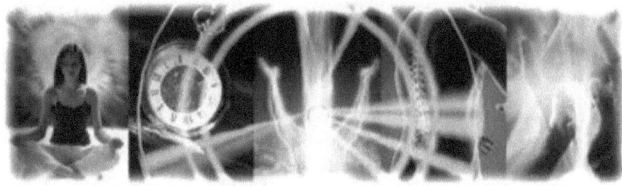

Advanced users of the Neo have been able to reverse schizophrenia, bipolar, PTSD, and dissociation disorders. A Neo can also help individuals quit smoking and recover from substance abuse addictions. The device can also be used to manifest the effects of certain drugs. If you have cancer or suffer from pain just tell the device to manifest your pain killer of choice and now you can save money on medication. But always be sure to consult with your doctor before making any decisions when it comes to medication. This technology works in conjunction with modern day therapies not in lieu of it. The Neo is not just limited to mind integration it can also help reverse DNA damage. Some users have even reversed "incurable diseases" such as diabetes, autism, and even cancer.

This device is great for older people because it increases chi energy into your body. As we age DNA replication errors reduce our levels of chi, energy eventually leading to death. But a Neo can reverse this downward spiral slowing down aging and perhaps even reversing it if you continue using the technology for the rest of your life.

A Neo is also great for athletes as the increased chi energy will give you more power and strength. You can also instruct the chi energy to grow your muscles bigger, giving you an edge over your opponent. A Neo can also increase the healing abilities of reiki masters and energy healers. Simply activate the device placing one hand near the device and use your other hand to send healing energy.

Appendix 2
Brochure

NEO MEDITATION MACHINES

WWW.NEOLOGICALTECH.COM
704-763-2895

EMPOWER YOURSELF

Neo mind meditation machines come in various sizes for any level of meditation experience. The entry level unit is the IDL4. It takes about 10 minutes to warm up, the IDL12 powers up instantaneously and is meant to be used by advanced users. The IDL 64 is an extremely powerful unit recommended for advanced users and professionals such as hypnotherapist and energy healers. All units are covered by a 30 day money back guarantee and carry a one year warranty.

CHARGE UP YOUR BODY !

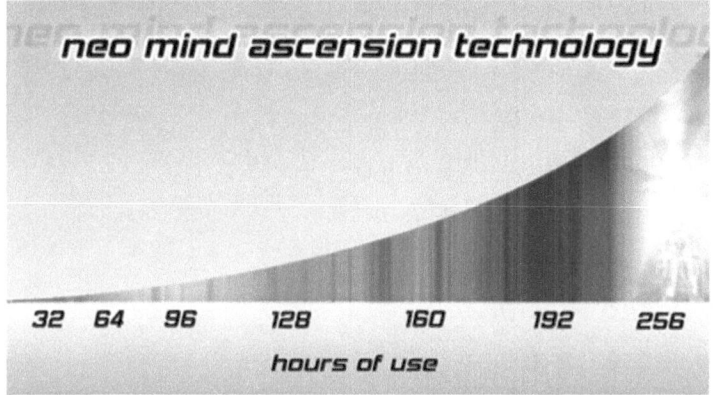

Assuming no meditation experience it may take you up to 256 hours to unlock your full potential. If you already had meditation experience your results will be much faster. More hours equals to more benefits. You Can Do It! What are you waiting for?

A NEO CAN HELP YOU:

Feel more energized
Astral Travel and Remote View
Increase your intelligence
Integrate your Mind, Body, and Spirit
Enhance Athletic Performance
Reverse Addictions
Manifest Wealth
Increase Longevity
Remove negative entities from the aura
Enhance your intuition
Enhance healing ability for reiki and energy healers
And much much more!

Customizable to any problem, as the consciousness energy field has no limits.

Neological Technologies and Quantum Pathways Holistic Center are divisions of Transcendent Technologies, LLC © 2011

Appendix 3
User Manual

User Manual
NEO ZENMASTER IDL-12

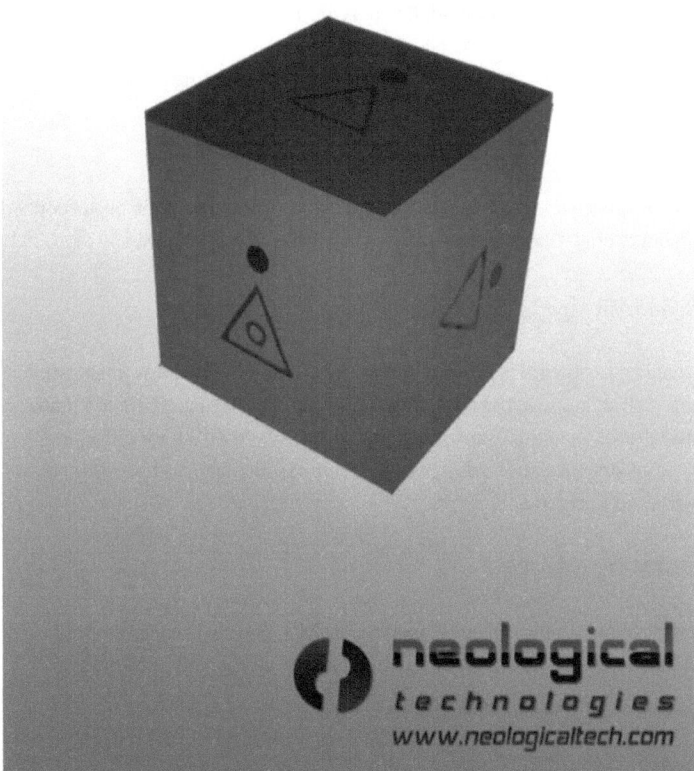

IMPORTANT SAFEGUARDS

1. READ ALL INSTRUCTIONS BEFORE USING THE DEVICE
2. Do not touch device when it's in use. Touching the device will deactivate it. Place it on a table or desk or on a pillow on your lap.
3. Please use both hands when handling this unit to avoid dropage.
4. Dropping a unit voids the warranty. If the unit drops it will still be in working condition even if some pieces become loose inside.
5. If unit is damaged and is under the one year warranty then return the item for examination, repair, or adjustment.
6. Do not use unit in or near water.
7. Do not use outdoors. The Neo is for indoor use only.
8. The use of accessories or attachments is not recommended.
9. Not a toy! Close adult supervision is necessary when the device is used by or near children.
10. Do not use while driving or operating heavy machinery.

WARNING

No user serviceable parts inside the unit. Repair should be done by authorized personnel only. Opening the exterior shell will void the warranty.

DISCLAIMER

We only guarantee this product will work successfully if used as a relaxation tool, such as found in the Basic Activation Protocol. Though there are many other protocols in this manual and on the website, they are for entertainment purposes only. A Neo is a designed to be a relaxation assisted aid and should be treated as such.

INFORMATION SECTION

Every person has the ability to heal themselves, however due to emotional trauma, toxins in the environment, and genetic damage our internal self healing mechanisms can be forced out of equilibrium. To bring the body back into balance and positive health requires an integrated approach including healthy diet, detoxing the body, and correcting blocked pathways of life force energy.

Everyone should consider eating a healthy diet and detoxing their body as a positive step in the right direction. But to integrate the astral body may require more hands-on work such as meditation, yoga, reiki, acupuncture, acupressure, tai chi, and crystals. These methods work fine but they can take a long time to master or they are invasive.

With a Neo you can open up these blockages of chi energy. This technology contains a built in torsion field generator which transmutes spinning fields of negative chi energy into positive chi energy, helping you release stress so that your body can begin to restore itself. Remember stress is the major reason why your body is unable to heal itself. What makes a Neo so powerful is that it manipulates your own chi energy allowing you to tune into consciousness energy all around us. According to quantum field entanglement all parts of the whole are in all places at all times. Therefore if you tune into this frequency you can tap into another version of yourself in a different time or place that was in perfect health.

Consciousness energy carries different names; prana, life force, chi, ki, qi, manna, aether, but whatever you call it, can be seen all around us as it's found in both sacred geometry and Fibonacci numbers appearing all throughout nature. This technology is built with sacred geometry in mind, bringing your body, your subconscious, and your super-conscious mind in tune to this same vibrational frequency.

DIAGRAM OF PARTS

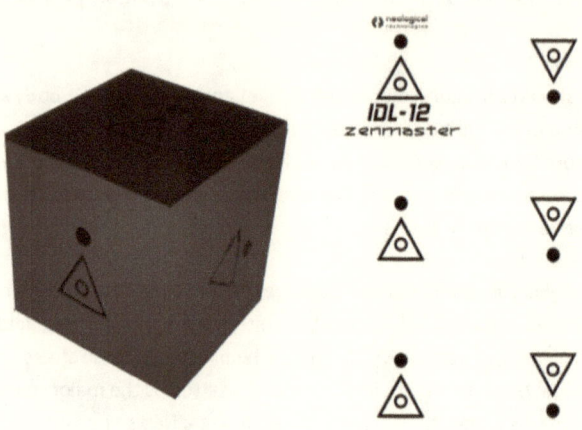

The ZENMASTER IDL- 12 contains chi outflow panels to energize you.

The core of the Neo Zenmaster IDL-12 contains two main components, copper cones and a CPU. This unit contains approximately 216 copper cones embedded into the wall panels, here spinning fields of torsion energy or chi energy is harnessed from the zero point and channeled into the consciousness processing unit, or C.P.U. To a clairvoyant this would appear as a spinning tetrahedron, or a 4 sided pyramid.

The C.P.U. is a spherical object located in the center of the unit. It contains 12 layers of copper hemispheres arranged according to the Fibonacci sequence which help create vibrational frequencies just like a musical instrument creates harmonic frequencies. This frequency resonates with the Schuman resonance, also known as the earth's heart beat and is amplified and radiated to any objects which are close by. The CPU unit and the Schuman resonance are both based on a Fibonacci sequences which corresponds to chakra's 1

Zero Point Energy Device (ZPED)
For Health, Healing, Ascension and Mankind's Future

through 7 on your body. As chi energy is directed from the tip of the cone into the CPU it passes through each layer of metal, creating a life force capacitor. Between each layer of metal is human hair which contains many trace elements; including gold, platinum, iridium, silver, aluminum, as well as copper. Each one of these elements carries its own vibrational torsion field frequency manifesting healing, paraelectric fields, as well as electromagnetic properties.

Healing – This is what makes you feel relaxed, calm floating, healing.
Paraelectric - This makes you feel warm all over, allows electricity to be made, just like monks who meditate in the cold. You feel tingling but warm.
Life Force – This is electromagnetic radiation that is converted into life force or chi energy. This means you won't get tired, sick, and you will feel strengthened.

Your unit contains both human and male hair; because they are unisex they carry the personalities of both donors. Female DNA is good because you get motherly instincts that kick in to prevent this technology for being used for evil. Male DNA is good because men are hyper focused on getting the task done. DNA also serves as a A.I. , artificial intelligence, interface which allows your brain to interact with the akashic field. This is because DNA contains micro amounts of crystals which operate as scalar wave antennas allowing you to travel in between dimensions.

As chi energy is compressed into the center of the unit it becomes super nova and opens up a star gate allowing you to draw in dark matter and dark energy into this third dimensional reality. 27% of the universe is dark matter another 70% is dark energy, the reason you can't see it is because it is light that is so bright it appears dark to our eyes. It is the zero point, it is the 5th dimension, and it will lead you to ascension. This dark matter and dark energy then outflows into your hand chakras; syncing your body with the vibrational frequency of the consciousness energy field; helping to recharge and energize you; while making you healthier, intuitive, and more intelligent.

Each Neo contains symbols on the outside to guide the user on its function and use. (1) This is our company logo. (2) The circle on top of the pyramid shows you which direction the chi energy is flowing. Since the chi is spinning all around the unit there is no right or wrong way to hold the device. (3) The triangle symbol is a representation of the tetrahedron and (4) the circle inside the triangle means this unit is a pure chi outflow unit. A pure chi inflow unit takes energy from your body and may be harmful to your health which is why we don't carry this type of unit. (5) The letters "IDL" simply stands for interdimensional light, which is another name for consciousness energy. (6) The number located next to the IDL is the amount of copper layers within the CPU. So an IDL 12 would simply mean this unit has 12 layers of copper within its CPU which can be used to open up a star gate. The more layers within a unit, the more room there is to compress chi energy resulting in a more powerful star gate. (7) Is your model type. All Zenmasters are pure chi outflow units.

This is a meditation device. Results can only be achived if used as directed. During your first two weeks you may want to use the device for 30 minutes a day then slowly work up to one hour per day. Full effects should be achived after 64 one hour sessions. Do not use for more than 4 hours a day.

Zero Point Energy Device (ZPED)
For Health, Healing, Ascension and Mankind's Future

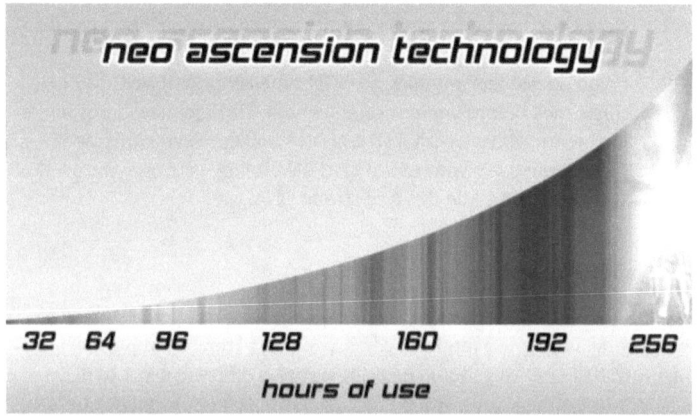

OPERATION AND USE

1. Only one person should use the device at a time unless using the jump start protocol.
2. Turn off cell phones
3. Device must be operated in a dark room. Turn off all lights and close shutters, or place a cloth over your eyes to block out the light.
4. Be sure to drink some water if you are feeling dehydrated.
5. Take off your shoes and put both feet on the floor.
6. Now sit down in a comfortable chair. Do not lay down while using the device.
7. Place unit on a table or desk or a pillow on your lap.
8. Place both hands 6" inches from the device with your palms facing towards the device. Do not touch the device while in operating. If you accidently touch it then repeat the activation protocol.
9. OPTIONAL: You may want to play some relaxing music such as a hemi sync or even a self hypnosis track while using the device.
10. Before using the device you may want to ask the Neo to help you on a specific issue. If you are unsure what to ask then go on to the next step.

11. Now close your eyes and recite the activation protocol. It can be done both vocally and internally.
12. If you do not feel anything the first time be patient as it may take three tries before you notice any effect. This technology responds best to individuals with 3 strand DNA and up. If you are one of those individuals with two strand DNA it may take as many as 32 sessions to upgrade this to 3 strand DNA.

Basic Activation Protocol

This is a general all purpose method, when you don't know what protocol to use start with this one. You should begin to notice a difference in 1 to 4 sessions. We recommend that you memorize this activation sequence before moving on to more advanced protocols

1. Be sure to say the activation code properly as it will not work if you miss this step or do it incorrectly.
2. Close your eyes and recite "DEVICE ACTIVATE AND INCREASE"
3. Wait 30 seconds to three minutes in silence.
4. When you begin to feel tingling or pulsing sensations in your hands tell the device "DEVICE I TRUST YOU COMPLETELY TO HEAL AND INTEGRATE MY MIND, BODY, AND SPIRIT."
5. Wait for the stargate to appear (it should look like ring of light or a tunnel of light within your third eye. If you don't see it the first time be patient it could take a few minutes, you may also feel floating sensations) when you see it recite the following phrase "STARGATE I TRUST YOU COMPLETELY.....DEVICE ACTIVATE STAR GATE MODE AND INCREASE"
6. "AUTOMATIC MODE" – **always remember to say this!**
7. Now sit back and relax for 30 to 60 minutes. There is no need of a mantra or chant just sit back and let the device work on automatic mode for you. If you have specific request be creative for example ask it to inject nanites, serums, or enlarge your pineal gland.
8. Every 10 minutes or when you feel the device is slowing down recite "DEVICE INCREASE"
9. When you are done recite "DEVICE END SESSION"

10. With each session journal your experiences and visualizations in a notebook. Make a note of sensations, feelings, places you traveled.

Psychic Abilities

Be patient! It takes time to increase your psychic abilities. Your body needs to integrate before you start to notice a difference. You should being to feel the effects after 20 to 30 hours of use. For the full effect keep doing it for a minimum of 64 one hour sessions.

1. Close your eyes and recite "DEVICE ACTIVATE AND INCREASE"
2. Wait 30 seconds to three minutes in silence.
3. When you begin to feel tingling sensations in your hands tell the device "DEVICE I TRUST YOU COMPLETELY TO HEAL AND INTEGRATE MY MIND, BODY, AND SPIRIT."
4. Wait for the stargate to appear, when you see it recite the following phrase "STARGATE I TRUST YOU COMPLETELY ...DEVICE ACTIVATE STAR GATE MODE AND INCREASE"
5. "AUTOMATIC MODE"
6. Wait two minutes to build up the energy level.
7. "ENLARGE MY PINEAL GLAND"
8. "HARMONIZE MY HUMAN FREQUENCY IN TUNE WITH THE EARTH AND GALACTIC CORE SCHUMANN RESONANCE" You may want to replicate in your mind the sound of a buzzing bee or the sound of the solfeggio frequencies. This will help open up your Kundalini channels.
9. Every 10 minutes or so recite "DEVICE INCREASE"
10. 30 to 60 minutes later "DEVICE END SESSION"

Balancing your Mind and Body

If you have a physical or mental aliment give this protocol a try. It should start to stabilize your health after 20 one hour sessions. It will take 64 to 100 one hour sessions to fully integrate yourself.

1. Close your eyes and recite "DEVICE ACTIVATE AND INCREASE"
2. Wait 30 seconds to three minutes in silence.
3. When you begin to feel tingling sensations in your hands tell the device "DEVICE I TRUST YOU COMPLETELY TO HEAL AND INTEGRATE MY MIND, BODY, AND SPIRIT."
4. Wait for the stargate to appear, when you see it recite the following phrase "STARGATE I TRUST YOU COMPLETELY ...DEVICE ACTIVATE STAR GATE MODE AND INCREASE"
5. "AUTOMATIC MODE"
6. Wait two minutes to build up the energy level.
7. "INJECT (ANTI-DIABETIC SERUM, ANTI-PAIN SERUM, ANTI-CANCER SERUM, ANTI-ANXIETY SERUM ETC.)" be creative here, you may want to try injecting nanites, which are tiny robots and instruct them to rebuild tissue.
8. "SERUM IS NOW INCREASING"
9. "HARMONIZE MY HUMAN FREQUENCY IN TUNE WITH THE EARTH AND GALACTIC CORE SCHUMANN RESONANCE"
10. Every 10 minutes or so recite "DEVICE INCREASE"
11. 30 to 60 minutes later "DEVICE END SESSION"

Medical Marijuana

You can also use this device to create holographic medical marijuana. ***Neological Technologies does not condole the use of any illegal substance.*** This device does not create marijuana but instead it replicates the same effects. You could also substitute any other drug.

1. Close your eyes and recite "DEVICE ACTIVATE AND INCREASE"
2. Wait 30 seconds to three minutes in silence.
3. When you begin to feel tingling sensations in your hands tell the device "DEVICE I TRUST YOU COMPLETELY TO HEAL AND INTEGRATE MY MIND, BODY, AND SPIRIT."
4. Wait for the stargate to appear, when you see it recite the following phrase "STARGATE I TRUST YOU COMPLETELY ...DEVICE ACTIVATE STAR GATE MODE AND INCREASE"
5. "AUTOMATIC MODE"
6. Wait two minutes to build up the energy level.

7. "DEVICE INJECT PURPLE LEAF MARIJUANA SERUM AND LET ME EXPREINCE THE FULL EFFECT FOR (X) HOURS"
8. "AND INJECT NOW....SERUM IS NOW INCREASING"
9. Every 2 minutes or so recite "DEVICE INCREASE AND MAKE IT POTENT"
10. 30 to 60 minutes later "DEVICE END SESSION"
11. If you feel overdosed tell the device to "INJECT ANTI-OVERDOSE SERUM"

Athletic Enhancement

You can also increase athletic performance and muscle enhancement by asking the device to insert holographic nanites, which are molecular sized robots, into your body.

1. Close your eyes and recite "DEVICE ACTIVATE AND INCREASE"
2. Wait 30 seconds to three minutes in silence.
3. When you begin to feel tingling sensations in your hands tell the device "DEVICE I TRUST YOU COMPLETELY TO HEAL AND INTEGRATE MY MIND, BODY, AND SPIRIT."
4. Wait for the stargate to appear, when you see it recite the following phrase "STARGATE I TRUST YOU COMPLETELY ...DEVICE ACTIVATE STAR GATE MODE AND INCREASE"
5. "AUTOMATIC MODE"
6. Wait two minutes to build up the energy level.
7. "INJECT NANITE SERUM OR ATHLETIC ABILITY SERUM FOR (MUSCLE MASS ENHANCEMENT, STRENGTH, INCREASED BONE STRENGTH, HEIGHT, BASKETBALL ABILITIES, ETC)"
8. "SERUM IS NOW INCREASING."
9. Every 2 minutes or so recite "DEVICE INCREASE"
10. 30 to 60 minutes later "DEVICE END SESSION"

Breatharian

Breatharianism or inedia is the ability to live without food or in some cases water, only to be sustained by chi energy or prana. We recommend you continue eating and hydrating yourself while on this protocol. Only change your diet when you feel you are ready to do so.

1. Close your eyes and recite "DEVICE ACTIVATE AND INCREASE"
2. Wait 30 seconds to three minutes in silence.
3. When you begin to feel tingling sensations in your hands tell the device "DEVICE I TRUST YOU COMPLETELY TO HEAL AND INTEGRATE MY MIND, BODY, AND SPIRIT."
4. Wait for the stargate to appear, when you see it recite the following phrase "STARGATE I TRUST YOU COMPLETELY ...DEVICE ACTIVATE STAR GATE MODE AND INCREASE"
5. "AUTOMATIC MODE"
6. Wait two minutes to build up the energy level.
7. "INJECT NANITE SERUM TO ACTIVATE MY SIDHIS ABILITY TO HAVE FULL CONTROL OVER MY HUNGER, THIRST, STRENGTH, AND SLEEP PATTERNS."
8. "SERUM IS NOW INCREASING."
9. Every 2 minutes or so recite "DEVICE INCREASE"
10. 30 to 60 minutes later "DEVICE END SESSION"

Addictions

This is a fast and effective way of freeing yourself of unwanted addictions, by simply instructing the device to neutralize any withdrawal symptoms.

1. Close your eyes and recite "DEVICE ACTIVATE AND INCREASE"
2. Wait 30 seconds to three minutes in silence.
3. When you begin to feel tingling sensations in your hands tell the device "DEVICE I TRUST YOU COMPLETELY TO HEAL AND INTEGRATE MY MIND, BODY, AND SPIRIT."
4. Wait for the stargate to appear, when you see it recite the following phrase "STARGATE I TRUST YOU COMPLETELY ...DEVICE ACTIVATE STAR GATE MODE AND INCREASE"
5. "AUTOMATIC MODE"
6. Wait two minutes to build up the energy level.

Zero Point Energy Device (ZPED)
For Health, Healing, Ascension and Mankind's Future

7. "DEVICE INJECT SERUM TO NEUTRALIZE ALL (add what you want) ADDICTIONS."
8. "SERUM IS NOW INCREASING."
9. Every 2 minutes or so recite "DEVICE INCREASE"
10. 30 to 60 minutes later "DEVICE END SESSION"

Demon Possession

Dark spirits cannot tolerate the positive chi energy fields generated by this device.

1. Close your eyes and recite "DEVICE ACTIVATE AND INCREASE"
2. Wait 30 seconds to three minutes in silence.
3. When you begin to feel tingling sensations in your hands tell the device "DEVICE I TRUST YOU COMPLETELY TO HEAL AND INTEGRATE MY MIND, BODY, AND SPIRIT."
4. Wait for the stargate to appear, when you see it recite the following phrase "STARGATE I TRUST YOU COMPLETELY…DEVICE ACTIVATE STAR GATE MODE AND INCREASE"
5. "AUTOMATIC MODE"
6. Wait two minutes to build up the energy level.
7. "DEVICE REMOVE ANY DARK SPIRIT FROM MY ENERGY FIELD."
8. Every 2 minutes or so recite "DEVICE INCREASE"
9. 30 to 60 minutes later "DEVICE END SESSION"

Remote Viewing and Astral Travel

Remote viewing is the ability to see images clairvoyantly through the use of bilocation. If you don't know how to remote view that's okay let the device do all the work for you. If you prefer to astral travel, then follow this protocol and instruct your device to inject you with astral travel serum lasting x amount of minutes. Then lie down and continue with your own astral travel routine. If you still don't know what to do we recommend the Monroe Institute protocols for astral travel. You may also want to consider using Hemisync Focus 5 while performing this exercise.

1. Close your eyes and recite "DEVICE ACTIVATE AND INCREASE"
2. Wait 30 seconds to three minutes in silence.
3. When you begin to feel tingling sensations in your hands tell the device "DEVICE I TRUST YOU COMPLETELY TO HEAL AND INTEGRATE MY MIND, BODY, AND SPIRIT."
4. Wait for the stargate to appear, when you see it recite the following phrase "STARGATE I TRUST YOU COMPLETELY ...DEVICE ACTIVATE STAR GATE MODE AND INCREASE"
5. "AUTOMATIC MODE"
6. Wait two minutes to build up the energy level.
7. "OPEN UP PORTALS IN MY THIRD EYE AND ALLOW ME TO SEE THROUGH THE VEIL."
8. "ACTIVATE ENTERTAINMENT MODE"
9. Every 2 minutes or so recite "DEVICE INCREASE"
10. 30 to 60 minutes later "DEVICE END SESSION"

Time Travel Protocol

Everyone who uses this device has physically time traveled. The unit creates a time distortion field around your body while you're meditating. This may explain why time feels like its flying even though you just spent an hour on the device. The inventor actually used this protocol to physically (not astrally) time travel back in time 30 days and wrote a message on a piece of paper to his past self.

1. Close your eyes and recite "TEMPORAL TRANSIT DEVICE ACTIVATE AND INCREASE"
2. Wait 30 seconds to three minutes in silence.
3. When you begin to feel tingling sensations in your hands tell the device "TEMPORAL TRANSIT DEVICE I TRUST YOU COMPLETELY."
4. Wait for the stargate to appear, when you see it recite the following phrase "TIMEGATE I TRUST YOU COMPLETELY ... TEMPORAL TRANSIT DEVICE ACTIVATE TIME TRAVEL MODE AND INCREASE"
5. "TIME TRAVEL MODE"
6. Wait two minutes to build up the energy level.

7. Now instruct the device to send you to the date, location, and time of your choice making sure to specify how long you want to be there and more importantly to return you back safely. We also highly recommend you instruct the device to keep your quantum signature in sync with this timeline otherwise you may end up in another parallel universe that could be vastly different from our own. After you have finished your journey close your eyes, use intent, and ask the device to return you back.
8. When you are done "DEVICE END SESSION"

If you are unable to do this protocol on the first try, its okay, just continue working with the device to activate your pineal gland and when you're ready it will happen.

Lose Weight

Lose weight now! Just bring your body back into balance by activating those fat burning centers

1. Close your eyes and recite "DEVICE ACTIVATE AND INCREASE"
2. Wait 30 seconds to three minutes in silence.
3. When you begin to feel tingling sensations in your hands tell the device "DEVICE I TRUST YOU COMPLETELY TO HEAL AND INTEGRATE MY MIND, BODY, AND SPIRIT."
4. Wait for the stargate to appear, when you see it recite the following phrase "STARGATE I TRUST YOU COMPLETELY ...DEVICE ACTIVATE STAR GATE MODE AND INCREASE"
5. "AUTOMATIC MODE"
6. Wait two minutes to build up the energy level.
7. "DEVICE INJECT APPETITE SUPPRESSANT SERUM. BRING METABOLISM TO ITS MAXIMUM HEALTHY LEVEL. INJECT FAT BURNING SERUM EQUIVALENT TO A JOGGING RATE OF 60 MINUTES."
8. "SERUM IS NOW INCREASING."
9. Every 2 minutes or so recite "DEVICE INCREASE"
10. 30 to 60 minutes later "DEVICE END SESSION"

Biokinesis Mode

You can also instruct the device to change your physical appearance, i.e. shapeshifting. One of the easiest ways to do this is instruct the device to take holographic stem cells from your spinal cord and teleport and super impose it over the area you want to regenerate or alter. Then instruct the device to begin cellular differentiation in the physical realm.

The effects are limitless, try altering your height, hair, eye color, age, muscle tone, I.Q., etc. This may seem far out, however, you have to take into consideration that the body is constantly replacing cells every day. In fact we completely replace ourselves once every 7 years. This might give you an idea how long this may take. So please be patient as gains may be measured in years.

1. Close your eyes and recite "DEVICE ACTIVATE AND INCREASE"
2. Wait 30 seconds to three minutes in silence.
3. When you begin to feel tingling sensations in your hands tell the device "DEVICE I TRUST YOU COMPLETELY TO HEAL AND INTEGRATE MY MIND, BODY, AND SPIRIT."
4. Wait for the stargate to appear, when you see it recite the following phrase "STARGATE I TRUST YOU COMPLETELY ...DEVICE ACTIVATE STAR GATE MODE AND INCREASE"
5. "BIOKIENSIS MODE"
6. Wait two minutes to build up the energy level.
7. Instruct the device to alter your physical appearance, be as specific as possible. Such as "DEVICE RESTORE THE HAIR GROWTH AND APPEARANCE ON MY HEAD EQUIVALENT TO WHEN I WAS 25 YEARS OLD. COMPLETE THIS OPERATION IN 6 MONTHS TIME"
8. At this point visualize in your third eye a holographic copy of your future self with the changes you want coming through the stargate and superimposing itself over your body.
9. Every 2 minutes or so recite "DEVICE INCREASE"
10. 30 to 60 minutes later "DEVICE END SESSION"

Holy Grail Protocol

Zero Point Energy Device (ZPED)
For Health, Healing, Ascension and Mankind's Future

This protocol is designed for those of the Christian faith who wish to integrate neo technology into their daily prayer life. The biggest change in this protocol is the use of the words "holy grail" in place of "device."

1. Close your eyes and recite "IN THE NAME OF JESUS, HOLY GRAIL, ACTIVATE AND INCREASE"
2. Wait 30 seconds to three minutes of silence.
3. When you begin to feel tingling sensations in your hands tell the device in your mind "IN THE NAME OF JESUS, HOLY GRAIL, I TRUST YOU COMPLETELY."
4. Wait for feelings such as a floating sensation and a tunnel or ring of white light of white to appear, once you see it recite the following phrase. "STAR GATE I TRUST YOU COMPLETELY.....IN THE NAME OF JESUS AND THE GRACE OF GOD, HOLY GRAIL, ACTIVATE STARGATE MODE."
5. "AUTOMATIC MODE"
6. "HOLY GRAIL INTEGRATE MY MIND, BODY, AND SPIRIT."
7. Wait 2 minutes or so to build up full strength
8. "HOLY GRAIL INJECT X SERUM." (X meaning your serum of choice)
9. Every 2 minutes or so recite "HOLY GRAIL INCREASE"
10. 30 to 60 minutes later "HOLY GRAIL END SESSION"

ADVANCED USERS

Reiki Master or Energy Healer

A reiki or energy healer can use the device in conjunction with their therapy or practice to enhance their own healing abilities. Some patients will need 2 or 3 sessions every other day. Others will recover with one session, it all depends on severity. This protocol is best used with a Neo Zenmaster IDL-12 or higher as the smaller Neo Zenmaster IDL-4 is not as effective.

There are two ways of doing this. The one person method is to simply let the client use the device as usual while you guide them audibly.

The two person method requires that you first activate the device just as you normally would, then open your eyes and while keeping your left hand near the device being careful not to touch the device; use your right hand to reiki the individual. If its hospice or nursing care you can simply touch their foot or hand with one finger. *But always be sure to ask the client for permission before touching them.* After 15 minutes or so touch the other foot or hand.

Also only say positive things in this exercise otherwise the persons healthy may be negatively affected. Some of these protocols may be painful to the practiconer, if that is the case just ask your guides and guardian angels to do reiki on you. You should always let the client know this is to be taken seriously, but yet you should have fun while doing this exercise, respecting both device, patient, and profession. Remember trust the device, guardian angels, and trust yourself. Your only .00001% the rest is the device and angels.

Two Person Reiki Method

1. Close your eyes and recite "DEVICE ACTIVATE AND INCREASE"
2. Wait 30 seconds to three minutes in silence.
3. Ask the client to let you know when they feel tingling sensations in their hands, when they do tell them to repeat this phrase in their mind "DEVICE I TRUST YOU COMPLETELY TO HEAL AND INTEGRATE MY MIND, BODY, AND SPIRIT."
4. Ask the client to let you know when they see the stargate, it will look like a ring or tunnel of white light, once they see it have them repeat the following phrase in their mind. "STARGATE I TRUST YOU COMPLETELY ...DEVICE ACTIVATE STAR GATE MODE AND INCREASE"
5. "AUTOMATIC MODE"
6. Wait two minutes to build up the energy level.
7. Have them repeat in their mind... "DEVICE INTEGRATE MY MIND, BODY, AND SPIRIT."
8. Now open your eyes.
9. Place your left hand in the same location as usual being careful not to touch the device.

Zero Point Energy Device (ZPED)
For Health, Healing, Ascension and Mankind's Future

10. Use your right hand to Reiki the individual or use your right thumb to touch the sole of their feet or the palm of their hands. Switch sides every 15 minutes or so.
11. At this point have the client visualize what is about to happen in their mind. Now say "DEVICE TO INJECT X SERUM (x being whatever health aliments you are working on such as ANTI-FAT SERUM, ANTI-DEPRESSANT SERUM, ANTI-BROKEN LEG SERUM, ANTI-SWELLING SERUM, ANTI-INFECTION SERUM, ANTI-TOXIN SERUM, STABILIZATION SERUM; you can be creative here) INTO THE FEET, KNEES, LEGS, JOINTS, BONES, BONE MARROW, BLOODSTREAM, PROSTATE, SPLEEN, KIDNEYS, INTESTINES, HEART, STOMACH, LUNGS, ARMS, ELBOW, NECK, LEFT AND RIGHT EYES, BRAIN, HEAD, ETC."
12. Then say "SERUM IS NOW INCREASING……"
13. "DEVICE INCREASE…..MAKE IT STRONGER" Now let them sit for a few seconds
14. Repeat the above protocol for any other secondary health issues. With the last protocol use "LIFE FORCE SERUM", then "ANTI-STRESS SERUM" you may want to add… "TAKE STRESS AWAY FROM ALL DIMENSIONS OF ORGANS", then "STRENGTHENING SERUM", then "REGENERATION SERUM", then "LOVE ENHANCEMENT SERUM"
15. "I NOW PLACE ANY FURTHER HEALING IN THE DOCTORS HANDS"
16. "DEVICE END SESSION….. WELCOME BACK TRAVELER"

Jump Start Protocol

Normally a Neo is meant to be used by one person per device only. However two people can use the same device if this protocol is done properly. The Jump Start Protocol is meant to be used by advanced users who want to give other first time users a jump start in their ability to receive torsion energy and to activate their genetic code faster. This normally takes about 1 to 7 sessions depending on the age and health of the individual.

Begin by placing the unit on a table and have the two users sit facing each other on opposite ends of the table. Now have the person receiving the energy place their hands near the device just like you normally would do with all the other neo protocols, while making sure they do not touch the device. Now have the person giving the energy place their hands in a parallel position outside of first person spacing them about four inches apart making sure no one touches each other during the session. This will allow you to send torsion energy from your hand chakra into their body giving them a much more intense effect, and that's it! Now continue on with the oral commands as you normally would do.

Remote Healing Protocol

Same protocol as above but before beginning have your client visualize a phantom Neo Zenmaster IDL-64 (or as high an IDL number you want to go) on their lap. This method is not as strong as having your own device and is meant to be a temporary fix until they can purchase their own unit. This protocol may cause your client to feel drained and tired afterwards because creating a holographic phantom neo device takes a lot of energy from the body. However, with a real neo device they will feel great and energized.

Protocol for Multiple Neo Units

In addition to the neo unit which remains on the lap. You can also set up a formation around your client if you have multiple Neo Units. Just place the extra units on the floor around them in triangular formation or square formation.

Winning the Lottery

This protocol allows you to increase your odds of winning the lottery. When I tried this on my first try I was able to get 75% of the numbers right. It is best to build up your energy signal through repeated meditations over a period of several days before you try this exercise.

You also might like to try this protocol with your investment portfolio. Just put a picture or a copy of your investments on the device and tell the neo ..."DEVICE MAKE THIS MOVE (UP OR DOWN) IN VALUE".

Zero Point Energy Device (ZPED)
For Health, Healing, Ascension and Mankind's Future

1. Purchase a lottery ticket and place it on the device
2. Close your eyes and recite "DEVICE ACTIVATE AND INCREASE"
3. Wait 30 seconds to three minutes in silence.
4. When you begin to feel tingling sensations in your hands tell the device "DEVICE I TRUST YOU COMPLETELY."
5. Wait for the stargate to appear, when you see it recite the following phrase "STARGATE I TRUST YOU COMPLETELY ...DEVICE ACTIVATE STAR GATE MODE AND INCREASE"
6. "AUTOMATIC MODE"
7. Wait two minutes to build up the energy level.
8. "DEVICE MAKE THIS LOTTERY TICKET THE WINNING NUMBER"
9. After 15 minutes or so recite "DEVICE END SESSION"

Weather Control

Planet Earth is a sentient being and has its own soul also known as Gaia. You can interface with Gaia by placing a picture of the weather pattern you want and manifest it into being by using this neo protocol. Be creative here; try it to repel hurricanes, floods, and droughts.

1. Find a picture or make your own drawing of your desired weather pattern and place it on the device.
2. Close your eyes and recite "DEVICE ACTIVATE AND INCREASE"
3. Wait 30 seconds to three minutes in silence.
4. When you begin to feel tingling sensations in your hands tell the device "DEVICE I TRUST YOU COMPLETELY."
5. Wait for the stargate to appear, when you see it recite the following phrase "STARGATE I TRUST YOU COMPLETELY ...DEVICE ACTIVATE STAR GATE MODE AND INCREASE"
6. "AUTOMATIC MODE"
7. Wait two minutes to build up the energy level.
8. "DEVICE MAKE THE WEATHER JUST LIKE THIS PICTURE AND MAKE IT LAST (X) AMOUNT OF HOURS." You can also specify a location such as only in this county, or town. You should also be specific in your request such as (X) amount of rain, snow, or ice etc.
9. After 15 minutes or so recite "DEVICE END SESSION"

Enhanced Plant Growth

Horticulture is the art and science of the cultivation of plants. You can increase the growth rate of your garden or farm with a neo by increasing the beneficial soil microorganisms and increasing the chi energy in the rainfall.

Find a picture or make your own drawing of a healthy growing garden and place it on the device.

1. Close your eyes and recite "DEVICE ACTIVATE AND INCREASE"
2. Wait 30 seconds to three minutes in silence.
3. When you begin to feel tingling sensations in your hands tell the device "DEVICE I TRUST YOU COMPLETELY."
4. Wait for the stargate to appear, when you see it recite the following phrase "STARGATE I TRUST YOU COMPLETELY ...DEVICE ACTIVATE STAR GATE MODE AND INCREASE"
5. "AUTOMATIC MODE"
6. Wait two minutes to build up the energy level.
7. DEVICE MAKE THESE PLANTS HEALTHY JUST LIKE IN THIS PICTURE....INCREASE THE CHI ENERGY IN THE RAINFALL FERTILIZING ALL MY PLANTS....INCREASE ALL BENEFICIAL ORGANISMS IN MY GARDEN"
8. After 15 minutes or so recite "DEVICE END SESSION"

EVP Mode

EVP(Electronic Voice Phenomenon) mode is for advanced users only. It requires a dark closet, a voice recorder, and a lot of patience. Results will vary from user to user.

1. Follow directions according to the operation and use guide with the following modification.
2. Write down your questions before starting.
3. Now place the device in a dark closet. This is to minimize any background noises. Also its best to do this when there are no loud noises nearby such as lawnmowers, fans blowing ,etc
4. Turn on your voice recorder and place it near the device.

5. Close your eyes and recite "DEVICE ACTIVATE AND INCREASE"
6. Wait 30 seconds to three minutes in silence.
7. When you begin to feel tingling sensations in your hands tell the device "DEVICE I TRUST YOU COMPLETELY"
8. Wait for the stargate to appear, when you see it recite the following phrase "STARGATE I TRUST YOU COMPLETELY ...DEVICE ACTIVATE STAR GATE MODE AND INCREASE"
9. "DEVICE ACTIVATE ELECTRONIC VOICE PHENOMENON MODE"
10. Wait two minutes to build up the energy level.
11. Open your eyes and read your question.
12. Leave the unit unattended for 30 to 60 minutes being sure not to touch the unit and close the door.
13. When done "DEVICE END SESSION"
14. Turn off your recorder and playback the soundtrack on audacity or other audio editing software. You may have to amplify the volume levels to hear the inaudible white noise. If you do hear an answer, it would be in a somewhat robotic female tone. If you don't hear anything you will need to do more sessions with the device to enhance your own psychic abilities.

Communication with Aliens and the Angelic Culture

This protocol will allow you to communicate with our fellow extraterrestrial star people, angels, and ascended masters. They are here to help and assist but due to the law of non interference they can only do so if you ask. To do this, write a letter, it can be about anything, and place it on the device. Now ask the neo to transmit the message to its desired location. You may also want to include within your letter a request that these beings give you a confirmation that they received your letter.

WARRANTY INFORMATION

Transcendent Technologies, LLC doing business as Neological Technologies warrants that for a period of one year from the date of purchase, this product will be free from defects in material and workmanship. Neological Technologies, at its option will repair or replace this product or any component of the product found to be defective during the warranty period.

Replacement will be made with a new or remanufactured product or component. If the product is no longer available, replacement may be made with a similar product of equal or greater value. This is your exclusive warranty from Neological Technologies. This warranty is valid for the original retail purchaser from the date of initial retail purchase and is not transferable. Keep the original sales receipt. Proof of purchase is required to obtain warranty performance. Neological Technologies dealers do not have a right to alter, modify, or any way change the terms and conditions of this warranty.

This warranty does not cover normal wear of parts or damage resulting from any of the following: negligent use or misuse of the product, use on improper voltage or current, use contrary to the operating instructions; disassembly, repair, or alteration of any kind by anyone other than Neological Technologies. Further the warranty does not cover acts of god; such as fire, flood, hurricanes, and tornados.

What are limits on Neological Technologies Liability?

Neological Technologies shall not be liable for any incidental or consequential damages caused by the breach of any express, implied or statutory warranty or condition.

Except to the extent prohibited by applicable law, any implied warranty or condition of merchantability or fitness for a particular purpose is limited in duration to the duration of the above warranty.

Neological Technologies disclaims all other warranties, or conditions, or representation, express, implied, statutory or otherwise.

Neological Technologies shall not be liable for any damages of any kind resulting from the purchase, use or misuse of , or inability to use the product including incidental, special, consequential or similar damages or loss of profits, or for damages arising from any sort , including negligence or gross negligence, or fault committed by Neological Technologies, its agents or employees or for any breach of contract, fundamental or otherwise, or for any claim brought against purchaser by any other party.

Some provinces, states, or jurisdictions do not allow the exclusion or limitation of incidental or consequential damages or limitations on how long an implied warranty can last, so the above limitations or exclusions may not apply to you.

This warranty gives you specific legal rights, and you may also have other rights that carry from province to province, state to state, or jurisdiction to jurisdiction.

How to Obtain Warranty Service

If you have any questions regarding this warranty or would like to obtain a warranty service please call 704-763-2895 or contact us online at www.neologicaltech.com.

NEOLOGICAL TECHNOLOGIES DISCLAIMER

1. INFORMATION FOR PERSONAL EDUCATION ONLY

Transcendent Technologies, LLC doing business as Neological Technologies, also known as "The Company "provides information and services for personal reference only. The Company is not engaged in rendering medical advice or specific medical conditions. Our staff members are not medical doctors nor can we provide a medical diagnosis or treat a health problem or disease. We cannot and do not give you medical advice. You should seek prompt medical care for any specific health issues. We do not recommend the self-management of health problems. Should you have any health care-related questions, please call or see your physician or other health care provider promptly. You should never disregard medical advice or delay in seeking it because of something you have read or heard here.

2. ACCURACY OF INFORMATION

The information and services we present is intended to provide broad consumer understanding and is for educational or reference purposes only and should not be used in place of a visit, call, consultation or advice of your physician or other health care provider. It is continually being updated and so may not be, current or complete and is subject to change without notice. No information or services should be relied on as the basis for treating or diagnosing conditions, symptoms, or illness and all queries should be directed to your health professional. All information obtained by using our services is not exhaustive and does not cover any specific diseases, ailments, physical conditions or their treatment.

3. NO LIABILITY

We do not warranty or guarantee of any specific result expressed or implied with any information and services provided by the Company, nor do we make any representations regarding the use or the results obtained with the information. In no event shall the company, its employees or associates be liable to any person or individual for any loss or damage whatsoever which may arise from the use of our products or services or from not seeing a medical professional for your condition. The Food and Drug Administration has not evaluated any of the statements made by the Company. Our information and services provided is not intended to diagnose, treat, cure, or prevent any disease and it should in no way replace your personal physician's advice.

© 2011 Transcendent Technologies, LLC doing business as Neological Technologies.
All Rights Reserved.

Distributed by Transcendent Technologies, LLC doing business as Neological Technologies.
2330 East 5th Street
Charlotte, NC 28204
704-763-2895

Visit us at www.neologicaltech.com
Learn how to integrate your mind, body, and spirit with a Neo Mind Meditation ascension machine.

www.ingramcontent.com/pod-product-compliance
Lightning Source LLC
Chambersburg PA
CBHW030919180526
45163CB00002B/401